1) IRON TRIANGLE p. 72
2) see p. p. 17

3) move to "VIABLE FIRMS" (p. 95) [illegible]

a) MECHANIZATION
b) LAND LEASING
c) CHANGES IN LABOR MARKET
d) size of land [illegible]

Q. fair under "CONTRACT"

[illegible] MARKETING of FIRMS / SUB-CONTRACTING, etc.
(see p. XXX p. 121)

TALKS ABOUT "TRIANGLE" — but
NOT [illegible]

JAPANESE AGRICULTURE UNDER SIEGE

JAPANESE AGRICULTURE UNDER SIEGE

The Political Economy of Agricultural Policies

Yujiro Hayami
Professor of International Economics
Aoyama-Gakuin University, Tokyo

St. Martin's Press New York

First published in the United States of America in 1988

Printed in Hong Kong

ISBN 0-312-01688-3

Library of Congress Cataloging-in-Publication Data
Hayami, Yujiro, 1932
Japanese agriculture under siege: the political economy of agricultural policies/by Yujiro Hayami.
p. cm.
Bibliography: p.
Includes index.
ISBN 0-312-01688-3: $30.00 (est.)
1. Agricultural price supports—Japan. 2. Agricultural and state–Japan. 3. Produce trade—Japan. 4. Agriculture and politics–Japan. I. Title.
HD2093.H387 1988
338.1′8—dc19 87-30761
CIP

Publication has been sponsored by the
Japan Foundation and the Suntory Foundation.

To Dr and Mrs Earl O. Heady

Contents

List of Figures ix
List of Tables x
Foreword by Malcolm Falkus xii
Preface xiii
Acknowledgements xv

1 The Roots of Agricultural Trade Friction 1
1.1 Growth in Agricultural Trade and Protection 2
1.2 The Agricultural Problem of the *Nouveaux Riches* 13
1.3 Plan of the Book 16

2 Changing Nature of the Agricultural Problem: A Historical Perspective 18
2.1 Agriculture in Economic Development 18
2.2 Overcoming the Food Problem 24
2.3 Era of the Poverty Problem 36
2.4 Towards the Dominance of the Agricultural Adjustment Problem 42

3 Structure of Agricultural Protection 51
3.1 Instruments of Agricultural Protection 51
3.2 Goals and Consequences of the Rice Policy 61
3.3 The Stalemate of Protection Policies 71

4 Failure of Structural Policies under the Agricultural Basic Law 74
4.1 The Bottleneck of Productivity Growth 74
4.2 An Unrealized Dream of Viable Farms 77
4.3 Dominance of Part-time Farming 83
4.4 Consequences of the Dominance of Part-time Farming 89

5 New Prospects for Structural Adjustment 95
5.1 Farm Mechanization and Scale Economies 95
5.2 Polarization in Progress 99
5.3 Labour Market Prospects 105

6 The Way to Lift the Siege 112
6.1 Towards an Open Trade System 112
6.2 The Mythology of Protectionism 121
6.3 Barriers to Policy Reorientation 125

Appendix A: A Model of the Political Market for Agricultural Protection 129

Appendix B: A Model of the Rice Policy in Japan 132

Bibliography 135

Index 139

List of Figures

2.1 Trends in total output, total input, total productivity, and labour and land productivities in Japanese agriculture (stock terms, five-year averages, semi-log scale) 28
2.2 Composition of agricultural production in Japan by commodities (nominal composition, current prices, five-year averages) 31
3.1 The structure of rice prices (000 yen per ton of brown rice) in Japan (average prices of the first and the second grades including packing costs, 1986) 65
3.2 Changes in the rice market and in the government financial burden due to rice price support programmes 68
5.1 Agricultural mechanization and economies of scale in Japanese agriculture 96
5.2 Comparison between average surplus from rice production per hectare of large farms (larger than 3 hectares) and average income from rice production per hectare of small farms (smaller than 3 hectares), deflated by the rural consumer price index (five-year moving averages) 101
5.3 Changes in average operational farm size made through purchases and leasing in Japan (excluding Hokkaido), by farm size class (1982) 103
5.4 Ratios of leased land to total operational land holdings by farm size class in Japan (1955–85) 104
5.5 Flow of the male labour force from agriculture to other sectors and its flow from other sectors into agriculture 109
5.6 Reasons why male workers leave non-agricultural jobs for agriculture and their future farming plans (1983) 111
A.1 Transition from the agricultural exploitation to the agricultural protection policy 130
B.1 Effects of the rice policy in Japan 133

List of Tables

1.1 Comparison of food self-sufficiency rates between Japan and other industrial countries (%) 3
1.2 Comparison of the nominal rates of agricultural protection between East Asian countries and other developed countries (1955–80) (%) 6
1.3 Nominal rates of agricultural protection for individual commodities in the EC and Japan (1955–80) 9
1.4 International comparison of the growth rates of labour productivity in agriculture and manufacturing (1960 (1958–62 averages) – 1980 (1978–82 averages)) 11
2.1 Agriculture in Japan's economic development (1885–1980) 20
2.2 Changes in agricultural and manufacturing commodity prices in Japan and the world 23
2.3 Distribution of farms by size of cultivated land area 27
2.4 Changes on direct tax burdens on the agricultural and the non-agricultural sectors in Japan 40
2.5 Changes in the allocation of government subsidies to the agricultural and the non-agricultural sectors 40
2.6 Estimates of nominal rates of protection on rice in Japan before the Second World War 42
3.1 Commodities subject to import quotas in Japan (as of May 1987) 53
3.2 Changes in the marketing margin corresponding to the abolition of the import quota (IQ) system for selected agricultural commodities in Japan 54
3.3 Structure of the national government budget for agriculture in Japan 57
3.4 Sources of finance for agricultural investments in Japan 58
3.5 Comparison of agricultural subsidies from central governments to Japan and EC member countries 60
3.6 Changes in the cost of the food control system in Japan 63
4.1 International comparisons in labour and land productivities,

and in land–labour ratios in agriculture, 1980 (1978–82 averages) 75
4.2 Changes in the number of new school-leavers in farm households who begin to engage in farming 79
4.3 Changes in the number and the shares of viable farms in Japanese agriculture 81
4.4 Changes in the agricultural labour force, number of farms and arable land area in Japan 82
4.5 Changes in distribution of gainful agricultural workers by sex and by age in Japan 85
4.6 Farm land prices and rates of return to land purchase in Japan 87
4.7 Distribution of farms that experienced a net increase in operational land-holding in Japan 88
4.8 Shares of various categories of farms in Japanese agriculture, 1985 89
4.9 Comparisons of income levels between farm household and urban worker households in Japan 92
5.1 Comparison of rice production costs in Japan by farm size, 1955–84 98
5.2 Changes in the number of male core farm workers aged over 15 and mainly engaged in work on their own farms 106
5,3 Distribution of part-time farm workers in Japan by non-farm economic activity in which engaged 107
6.1 Changes in food consumption in Japan 117
6.2 Comparison of protein intake level between Japan and other industrial countries 117

Foreword

As late as the mid-1950s some two-fifths of Japan's occupied population were involved in agricultural production (mostly peasant rice farming) and this sector contributed more than one-fifth of gross domestic product. Today the proportion so employed has shrunk to well under 10 per cent, while the contribution to total output is tiny. In this short span of little more than thirty years Japan has undergone a dramatic economic transformation which has taken her from a relatively poor and backward economy to the world's most powerful industrial economy outside the USA. In this process of change the role of agriculture has been a major one, both for its direct contribution to the transformation and also for its social and political place within Japan. Despite agriculture's present minor position in Japan's economy, it remains disproportionately strong in its political influence. Agricultural protection is also a key issue, affecting not only living standards and social structure within Japan, but also Japan's relations with Third World countries and the broader question of Japan's balance of payments with the rest of the world. Agriculture in Japan stands at a crossroads.

Professor Hayami, one of Japan's leading agricultural economists, has written a brilliant and authoritative book which explores fully the recent changes and present dilemmas confronting Japanese agriculture. This work addresses a major subject and will acquaint Western readers both with Professor Hayami's own views and research and also the results of much current Japanese scholarship in this area.

MALCOLM FALKUS

Preface

Visitors to Japan from abroad are often shocked by exorbitant food prices, such as a modest steak dinner costing some £50 sterling and a glass of fresh orange juice costing more than a cognac. While the price list in luxurious western-style hotels where they stay is by no means representative, there is no doubt that food prices are higher in Japan than in most other developed countries. How can the Japanese tolerate such high food prices? Why do they continue to resist opening doors to the imports of agricultural products, even at the risk of jeopardizing the free trade regime on which the present prosperity of Japan has been built? This book represents an attempt to answer these questions by a search into interactions between economy and politics. It intends to illustrate a polar case of the political economy of agricultural policies in affluent societies in contrast with another polar case in low-income economies, as described in Robert Bates's classic *Markets and States in Tropical Africa* (University of California Press, 1981).

The materials in this volume were prepared originally for an additional chapter to an as yet unpublished revision of my previous book, *A Century of Agricultural Growth in Japan* (University of Tokyo Press and University of Minnesota Press, 1975) that analyses the long-term growth of Japanese agriculture. Soon, I found it impossible to condense the materials into a single chapter, and decided to put them together in a new book focusing on current situations and policies. Detailed historical accounts of agricultural development in modern Japan should therefore be sought in the previous publication, even though they are summarized briefly in this volume.

Three earlier projects have paved the way for the preparation of this book. First, the joint project by Kym Anderson and myself on agricultural protection in East Asia, supported by the Japan–Australia Research Centre in Canberra and the Forum for Policy Innovation in Tokyo, which culminated in the publication of *The Political Economy of Agricultural Protection* by Anderson, Hayami *et al.* (Allen & Unwin, 1986). Second, my paper titled 'Agricultural Protectionism in the Industrialized World: the Case of Japan', presented at a conference on agricultural protectionism held in Honolulu in February 1986 under the sponsorship of the National Center for Food and Agricultural Policy in Washington, DC. Third, a textbook in Japanese, which was published with the title, *Nogyo Keizairon* (Agricultural Economics) (Iwanami Shoten, 1986) that includes much of the material in this volume, although it is organized in a very different manner.

I owe much to comments and suggestions from Kym Anderson, Fred Sanderson and other participants in the first and the second projects, as well as from David Feeny, Koichi Hamada, James Houck, Kenzo Henmi, Toshiyuki Kako, John Mellor, Mancur Olson, Vernon Ruttan, C. L. J. van der Meer, Saburo Yamada and Yasukichi Yasuba at various stages of research. It was Gustav Ranis who originally suggested to me the title of this book. Editorial suggestions from Malcolm Falkus were also highly appreciated. My greatest indebtedness is to three young friends, Masayoshi Honma, Toshihiko Kawagoe and Keijiro Otsuka, who assisted me in data collection and analysis for the preceding three projects as well as for the preparation of this volume.

I acknowledge with gratitude the permission from Allen & Unwin, the National Center for Food and Agricultural Policy and Iwanami Shoten for extensive use of material from the earlier publications. Financial support from the Japan Foundation and the Suntory Foundation for this publication are also gratefully acknowledged.

This book is dedicated to Dr and Mrs Earl O. Heady with my warm memory of their kind assistance and guidance, in both academic and personal spheres, when I was a lonely foreign student in Ames, Iowa, some thirty years ago.

YUJIRO HAYAMI

1 The Roots of Agricultural Trade Friction

Whenever economic friction arises between Japan and its trade partners, criticism is invariably made of the closed nature of Japanese agriculture. Import quotas on beef and oranges have been publicized as a symbol of unfair trade practice in Japan. Traditional attacks on Japan's protectionism from Australasia and North America have been joined recently by the Third World. Tariffs on labour-intensive, high value-added commodities such as boneless chicken and plywood are now spoiling the image of Japan in the mind of people in Asia, especially ASEAN (the Association of Southeast Asian Nations). Indeed, Japanese agriculture is under siege.

Inevitably the external pressure for agricultural trade liberalization increases as Japan's trade surplus continues to be large. It is feared that the trade friction might grow to such an extent as to jeopardize the present international order of free trade, on which Japan's economic success since the Second World War has been based. Why does Japan adhere to agricultural protectionism even at the risk of sacrificing the very source of its economic prosperity? It is the purpose of this book to investigate the roots of agricultural trade friction embedded in the economic and social structure of Japan. The analysis is focused on the problem of agricultural adjustment in rapid economic development based mainly on the borrowing of industrial technology. By nature, the problem is of the political economy, involving dynamic interactions among vested-interest groups. This introductory chapter attempts to give a perspective on the problem with the aid of broad international comparisons.

1.1 GROWTH IN AGRICULTURAL TRADE AND PROTECTION

Is the Japanese market so strongly protected from the competition of foreign agricultural producers as to warrant worldwide blame? Agricultural protectionism is not unique to Japan. Even the United States, taking leadership in the formation of the postwar liberal trade regime under the General Agreement on Tariffs and Trade (GATT), openly excepted some of the agricultural commodities from trade liberalization, in violation of the ideal of GATT itself. The formation and subsequent enlargement of the European Economic Community (EC) has severely limited opportunities of agricultural trade, as have the policies of numerous other European countries. During the past two decades the example of Western Europe in adopting increasingly protectionist agricultural policies has been followed first by Japan and then by other newly industrializing East Asian economies (Korea and Taiwan) that previously had a policy regime which – as in many less developed countries – discouraged agricultural production. Is there a unique factor in Japan that makes agricultural trade a focus for international economic friction?

1.1.1 Growth in Imports of Agricultural Commodities

Recurrent trade friction may give an impression that Japan has closed its door to the imports of foreign agricultural commodities. On the contrary, Japan has been one of the biggest contributors to expansion in world agricultural trade during the post-Second World War era. In fact, the real value of imports of agricultural commodities to Japan increased from 1960 to 1985 at an average rate of 13 per cent per year, more than three times faster than the world total. Meanwhile, the import quantities of major tradeable food items increased from 2.6 to 5.2 million tons in wheat, from 1.1 to 4.9 million tons in soybean and from 1.4 to 21.9 million tons in feed grains. Japan's shares of world imports amount to about 5 per cent in wheat and nearly 20 per cent in both soybean and feed grains. Japan stands now as one of the two largest net food importers in the world, rivalled only by the Soviet Union.

For the past several decades, Japanese agricultural imports have increased not only relative to the world total but also relative to domestic production. As a result, Japan's self-sufficiency in foods declined, as shown in Table 1.1. There has been a precipitous drop in

TABLE 1.1 Comparison of food self-sufficiency rates between Japan and other industrial countries (%)

	Japan			*1982*					
	1960	*1975*	*1985*	*USA*	*UK*	*France*	*Germany (FR)*	*Italy*	*Netherlands*
Grains	83	43	34	183	111	179	95	89	31
Food grains	91	76	74	312	109	208	104	99	59
Rice	102	110	107						
Wheat	39	4	14	320	111	216	109	94	62
Coarse grains	66	2	2	161	114	156	90	79	14
Pulses	44	9	8	147	51	119	16	98	9
Vegetables	100	99	95	102	66	92	36	122	255
Fruits	100	84	76	90	18	69	51	117	28
Dairy products	89	82	89	99	96	116	127	67	183
Eggs	101	97	98	103	99	105	71	95	298
Meat	91	77	81	98	78	100	89	75	213
Total final food Consumption	**91**	**76**	**73**						

Source: Japanese Ministry of Agriculture, Forestry and Fisheries (JMAFF), *Shokuryo Jikyu Hyo* (Food Balance Sheets); OECD, *Food Consumption Statistics*.

self-sufficiency in dry field crops such as wheat, feed grains and pulses that are heavily land-intensive and with which Japanese production is at a comparative disadvantage. Although more than complete self-sufficiency has been maintained for rice under the protection of the Food Control System (see Chapter 3), the average self-sufficiency rate for all grains was only 34 per cent in 1985, the lowest among major industrial countries. In this respect, the agricultural production and trade performance of Japan has been diametrically different from that of the EC. Under the protection of the Common Agricultural Policy (CAP), domestic agricultural production in the EC increased faster than domestic consumption, resulting in marked improvements in food self-sufficiency.

1.1.2 Growth in Agricultural Protection

A peculiar feature of agricultural trade in Japan is that sharp increases in imports and declines in self-sufficiency have been accompanied by rapid growth in the protection of domestic agriculture. The unique aspects of agricultural protection in Japan can best be understood by comparison with other industrialized and industrializing countries. A measure used for comparison is the average nominal rate of protection (NRP), calculated by subtracting the value of agricultural output in international prices from the value of agricultural output in domestic prices and dividing the remainder by the value of agricultural output in international prices; this is equivalent to the weighted average of the NRPs of individual commodities using their shares in the total output value at border prices as weights. Commodities covered in the average NRP calculation include twelve tradeable commodities (see footnote to Table 1.2 below) which currently account for about 60–70 per cent of the value of total agricultural output in cash of countries under study.

Table 1.2 summarizes the average NRPs estimated for fifteen industrial and newly industrializing countries by comparing producer and border (import cif) prices from seven of the years between 1955 and 1984. Producer prices are used because they include the effects not only of border protection but also of more direct agricultural support policies such as deficiency payments. However, the use of producer prices leads to an underestimation of protection to the extent that there are costs of marketing from the farm gate to a point in the marketing chain equivalent to the internationally traded product. This bias is obvious in the case of the food-exporting countries such as Australia and the US, for which the estimates of nominal protection rates are negative in some years when in fact no policy was exercised to exploit agriculture – or, rather, modest protective policies were adopted. However, in so far as this bias is similar across countries and over time, it does not present a serious problem for the purpose of making broad comparisons.

The first point worth noting about Table 1.2 is the high level of agricultural protection in Japan and two other newly industrializing countries (NICs) in East Asia. In 1984, the average NRPs of Japan (102 per cent), Korea (137 per cent) and Taiwan (43 per cent) were considerably higher than the EC average (22 per cent) and comparable only with those of Switzerland and Sweden which are known for their high level of agricultural protection for the purpose of national security and environmental conservation. 1984 was the year in which US dollar appreciated abnormally relative to the currencies of other industrial countries, due partly to the money supply control of the Reagan administration. This had the effect of lowering the domestic prices of agricultural commodities in other countries relative to the import prices in dollar terms. In order to assess the impact of the depreciation of the dollar since the late 1985, the NRPs of 1984 are recalculated by substituting the September 1986 for the 1984 exchange rates. The results of recalculation, as shown in parentheses, indicate that the average NRPs of Western Europe and Japan increased sharply corresponding to the adjustments in exchange rates, while that of Korea (whose currency is linked with US dollar) did not increase.

In any case, there is no doubt that the level of agricultural protection in Japan, as measured by the average NRPs, is among the highest in the world in recent years. With that level of protection, it is not surprising to see strong external pressure mounting on Japan for agricultural trade liberalization, given her large trade surplus.

However, it is noteworthy that the high rates of agricultural protection in Japan and two East Asian NICs are relatively recent phenomena. The average NRP of Japan in 1955 was 18 per cent, only half the EC average of 35 per cent. It then rose rapidly, reaching the EC level in 1960 and the Swiss level in 1965. This was the period when Japan's economic growth was especially fast. More dramatic were the cases of Korea and Taiwan. Before the mid-1960s, when the spurt of their industrial development began, the average NRPs were negative,

TABLE 1.2 Comparison of the nominal rates of agricultural protection between East Asian countries and 12 other developed countries, 1955–80 (%)[a]

	1955	*1960*	*65*	*1970*	*75*	*1980*	*84*	*(86)*[b]
East Asia:								
Japan	18	41	69	74	76	85	102	(210)
Korea	–46	–15	–4	29	30	117	137	(117)
Taiwan	–17	–3	–1	2	20	52	43	(na)
EC:								
Denmark	5	3	5	17	19	25	12	(46)
France	33	26	30	47	29	30	25	(75)
Germany, FR	35	48	55	50	39	44	49	(86)
Italy	47	50	66	69	38	57	20	(67)
Netherlands	14	21	35	41	32	27	15	(58)
UK	40	37	20	27	6	35	15	(54)
Average[c]	**35**	**37**	**45**	**52**	**29**	**38**	**22**	**(63)**

Non-EC:								
Sweden	34	44	50	65	43	59	36	(63)
Switzerland	60	64	73	96	96	126	153	(260)
Food exporters:								
Australia	5	7	5	7	–5	–2	na	(na)
Canada	0	4	2	5	7	–3	–3	(–9)
New Zealand	na	2	0	–5	–4	2	na	(na)
United States	2	1	9	11	4	0	6	(6)

Notes:
a Defined as the percentage by which the producer price exceeds the border price. The estimates shown are the weighted averages for 12 commodities, using production valued at border prices as weights. The 12 commodities are rice, wheat, barley, corn, oats, rye, beef, pork, chicken, eggs, milk and sugar.
b Calculated by applying the exchange rates of September 1986 to the 1984 prices.
c Weighted average for all six countries shown after 1970, but excluding Denmark and the UK for earlier years.

Source: Kym Anderson and Yujiro Hayami, *et al*. *The Political Economy of Agricultural Protection: East Asia in International Perspective* (Sydney, London and Boston: Allen & Unwin, 1986) p. 26; Masayoshi Honma, 'Kokusaiteki Kanten Karamita Nihon Nogyo no Hogosuijun' (Agricultural Protection Level of Japan in an International Perspective), paper presented at the Modern Economics Mini-Conference, held at the Hitotsubashi University Institute for Economic Research, 31 January 1987.

reflecting the practice of agricultural production policies common to low-income countries. During the 1970s their protection levels rose sharply, and Korea caught up Japan by 1980.

The average NRPs increased from 1955 to 1970 not only in Japan but also in industrial countries in general. This was the period when the international terms of trade turned against agriculture under the pressure of accumulated surpluses of grains in the US and other major exporters. On the other hand, precipitous drops in the average NRPs of industrial countries were experienced from 1970 to 1975, corresponding to sharp increases in agricultural prices relative to manufactured commodity prices during the so-called 'World Food Crisis'. The inverse correlation between NRP and the terms of trade for agriculture implies that industrial countries tried to insulate the domestic market from fluctuations in agricultural commodity prices in international market, with the logical consequence of creating domestic price stability at the expense of international instability (Johnson, 1973). However, even during the World Food Crisis period, Japan continued to increase its agricultural protection level, together with Korea and Taiwan.

1.1.3 Protection by Commodity

What are the major characteristics of agricultural protection in Japan on a commodity-by-commodity basis? Table 1.3 compares the NRPs of individual agricultural commodities between the EC and Japan. A distinctive characteristic of Japan is that the level of protection on grains is higher than on livestock products, while the reverse is the case with the EC. The protection on grains in Japan has a two-tier structure.It is extremely high for food grains (especially rice), but feed grains such as maize are imported virtually with no trade barrier, to such an extent that no appreciable amount is produced domestically and therefore no domestic producer price was recorded.

Such a strong bias in favour of food grains in Japan's agricultural protection seems to be explained by the traditional lack of substitutability between rice and feed grains that made it possible to increase the price support on rice and, at the same time, to import feed grains without protection (Hemmi, 1970). In European countries, it would have been difficult to maintain the prices of food grains at a high level if cheap feed grains were imported freely, because their food grains (such as wheat and rye) are highly substitutable for feed grains. Likewise, it would have been difficult for European countries to raise

TABLE 1.3 Nominal rates of agricultural protection for individual commodities in the EC and Japan, 1955–80 (%)

	EC						Japan					
	1955	*1960*	*1970*	*1980*	*84*	*(86)*	*1955*	*1960*	*1970*	*1980*	*84*	*(86)*
Grains:												
Rice	17	39	40	44	10	(38)	24	47	135	192	235	(414)
Wheat	46	36	54	18	–10	(20)	31	51	134	261	318	(541)
Barley	31	26	67	23	–8	(24)	24	52	158	307	363	(611)
Corn	20	14	23	38	–9	(17)	na	na	na	na	na	(na)
Oats	0	5	25	26	–11	(20)	na	na	na	na	na	(na)
Rye	47	44	46	32	6	(45)	na	na	na	na	na	(na)
Average	**33**	**29**	**47**	**23**	**–9**	**(21)**	**24**	**48**	**135**	**196**	**239**	**(420)**
Livestock:												
Beef	71	61	75	93	111	(181)	39	84	108	100	103	(211)
Pork	29	31	21	13	7	(44)	2	97	–9	17	21	(86)
Chicken	78	52	22	13	21	(60)	–52	19	18	23	9	(67)
Eggs	16	26	15	5	3	(38)	–19	–7	–9	–1	–7	(42)
Milk	16	29	86	53	39	(87)	4	5	212	186	185	(338)
Average	**34**	**37**	**52**	**42**	**38**	**(85)**	**–8**	**22**	**24**	**40**	**41**	**(116)**
Other:												
Sugar beet	101	100	91	40	–2	(31)	na	na	214	141	186	(339)
All commodities:	**35**	**37**	**52**	**38**	**22**	**(63)**	**18**	**41**	**74**	**85**	**102**	**(210)**

Source: JMAFF, *Shokuryo Jikyu Hyo* (Food Balance Sheets); OECD, *Food Consumption Statistics*.

the level of support for food grains too high because it would have necessarily resulted in high feed costs to domestic livestock producers.

In Japan, the lack of substitution between rice and feed grains enabled the simultaneous achievement of very high support on domestic food grains and low feed costs to domestic livestock producers. Yet past trends show that the NRP of livestock products in Japan increased faster than that of grains; in 1955 it was negative (considered to be zero if adjusted for marketing margins from the farm-gate to the wholesale level) and obviously much lower in Japan than the EC's 33 per cent, but it rose rapidly, reaching almost a par with the EC in 1980. Considering the lower rate of protection of feed grains, the effective rate of protection of livestock products in Japan in 1980 may well have been higher than in the EC.

Comparison of Table 1.3 with Table 1.1 reveals a tendency for the commodity structure of agricultural production to be determined by protection policy as well as comparative advantage. Japan is characterized by the scarce endowment of land relative to labour and capital. Comparative advantage is therefore lower in land-intensive crops such as grains and soybeans than in labour- and capital-intensive commodities such as vegetables, pigs and poultry. However, among grains more than complete self-sufficiency of rice has been maintained by very strong protection. Also, among livestock products beef and dairy products have been heavily protected. Further, among chickens a higher tariff rate has been applied to boneless chicken than boned chicken in carcase.

Such structure of protection clearly 'indicates a basic strategy of Japan's current food policy: to support a high self-sufficiency in consumer-ready food, and to rely heavily on foreign sources for raw-material types of agricultural products. This approach resembles Japan's industrial strategy of importing raw materials and then processing them, while blocking higher value-added imports to protect domestic manufacturers, especially in early stages of development'. (Reich, Endo and Timmer, 1986, p. 157). Naturally this structure creates serious friction with Australia on beef, the EC and New Zealand on dairy products, and ASEAN, China and Korea on labour-intensive, high value-added commodities such as boneless chicken, plywood and silk. With this structure Japanese agriculture is largely complementary with the exporters of feed grains, especially the US. Yet, Japan has not escaped friction with the US with respect to high-quality beef, oranges, tomato products and rice.

1.1.4 Causes and Effects of Changing Comparative Advantage

Underlying the increase in agricultural protection and the decrease in food self-sufficiency has been the sharp decline in comparative advantage in agriculture. The high rate of economic growth in Japan that lasted approximately from the mid-1950s to the first Oil Crisis in 1973 was led by the extremely rapid growth in industrial productivity. Table 1.4 compares the growth rates of labour productivities in the agricultural versus the manufacturing sectors from 1960 to 1980 among developed and developing countries for which data are available. During this period, labour productivity in agriculture in Japan increased at an average rate of 5.3 per cent per year in real terms, which was about the same speed as in other industrial countries. However, labour productivity in the manufacturing sector of Japan increased at a rate of 6.7 per cent per year, which was considerably higher than in the US and many European countries. As a result,

TABLE 1.4 International comparison of the growth rates of labour productivity in agriculture and manufacturing, 1960 (1958–62 averages) – 80 (1978–82 averages)

	Labour productivity growth rate (%/year)[a]		
	Agriculture (1)	*Manufacturing (2)*	*(1) – (2) = (3)*
Developed countries:			
US	6.3	3.2	3.1
UK	5.5	2.6	2.9
France	6.4	4.2	2.2
Germany (FR)	7.7	4.1	3.6
Japan	5.3	6.7	–1.4
Developing countries:			
Korea	4.0	7.5	–3.5
Philippines	3.2	3.5[b]	–0.3
India	1.3	2.1	–0.9

Notes:
a Calculated from the ratios of the real output index to the employment index.
b Growth rate 1960–75.

Source: FAO, *Production Yearbook*; UN, Yearbook of *Industrial Statistics*; ILO, Yearbook of *Labour Statistics*; OECD, *Labour Force Statistics*.

comparative advantage in agriculture was substantially lost in Japan relative to other industrial countries.

The decline in comparative advantage in agriculture, coupled with rapid increases in the consumption of high-valued foodstuffs such as meat and milk corresponding to *per capita* income rises, resulted in sharp increases in food and feed imports. At the same time, it created a demand for agricultural protection. If adjustments in intersectoral resource allocation corresponding to the rapid shift in comparative advantage from agriculture to industry had been left to market mechanisms, the cost of adjustment that rural people would have had to shoulder in such forms as rural–urban income disparity and depopulation in rural communications would have risen to a socially intolerable level. In order to shift a part of the industrial adjustment cost to the general public, farmers organized political lobbying for protection.

The decline in comparative advantage in agriculture due to rapid growth in industrial productivity was even sharper in Korea than in Japan. Korea's industrial development is similar to Japan's at the point that it is based on 'borrowed technology' in the Gerschenkron (1962) sense. The industrial growth rate tends to be higher for countries that begin systematic technology borrowing in later years, because the gap between the frontier technology in advanced industrial countries and the technology used in developing countries has become larger in more recent years. The rapid growth in agricultural protection in Japan during the 1960s and in Korea (and Taiwan) during the 1970s seems to reflect a general tendency that the loss of comparative advantage in agriculture is especially rapid and, therefore, the potential cost of interindustrial adjustment to be shouldered by farm producers becomes very large at the NICs stage of economic development based on industrial technology borrowing.

Even in lower-income countries such as the Philippines and India which have not yet entered the NICs stage, labour productivity in manufacturing increased faster than in agriculture, implying that their comparative advantage in agriculture declined relative to industrial countries. This is consistent with the observation that developing countries today which used to be net food exporters as a whole have become net importers since the Second World War, and their import margin has been rising. The shift of comparative advantage in agriculture from developing to developed countries reflects a shift in the basis of comparative advantage from natural resource endowments to scientific and engineering capacity (Hayami and Ruttan,

1985, Chapter 12). The fact that labour productivity in manufacturing increased faster than in agriculture in developing countries seems to show that it is easier for them to borrow industrial technology than to borrow agricultural technology; the direct importation of foreign agricultural technology is more difficult because it is more strongly constrained by climatic and ecological conditions than industrial technology.

The same economic force that has raised agricultural protection in Japan has increased foreign protests against it. The rapid shift in comparative advantage from agriculture to manufacturing has resulted in a surge of industrial exports from Japan; this has created severe industrial adjustment problems in the countries whose market has been penetrated by Japanese manufacturing products.

In those countries, it is only natural that strong demands have been made for Japan to liberalize agricultural trade, as well as demands for the protection of domestic manufacture. The former demand is often made by those interested in promoting the latter demand (as a means of publicizing Japan's unfairness) in order to rationalize industrial protectionism. So much the more dangerous is agricultural trade friction, therefore, which might become a factor in the destruction of the free trade regime.

1.2 THE AGRICULTURAL PROBLEM OF THE *NOUVEAUX RICHES*

It should now be clear that the high level of agricultural protection in Japan is rooted in the rapid decline in comparative advantage in agriculture mainly due to extremely rapid growth in industrial productivity based on borrowed technology. Why, then, are other industrial countries also adopting agricultural protection policies despite their improvements in comparative advantage in agriculture? And why are the developing countries that are losing agriculture's comparative advantage relative to industrial countries not adopting agricultural protection policies? Instead, it is common for developing countries to exploit agriculture by various means such as export taxes and overvalued exchange rates so that their domestic prices of agricultural commodities are depressed below international prices (Johnson, 1973; Schultz, 1978; Bale and Lutz, 1981; World Bank, 1986). Such contrast indicates that agricultural protection in Japan has a root that is common to other industrial countries.

1.2.1 Structure of the Agricultural Adjustment Problem

In general, agricultural protection in industrial countries is considered a manifestation of the 'agricultural adjustment problem'. This can be defined as the difficulty of reallocating resources (especially labour) from the agricultural to the non-agricultural sector corresponding to a relative contraction in the demand for agricultural products in the course of economic development – at a sufficiently rapid speed to prevent a decline in the rate of return to labour used in agriculture relative to that of the rest of the economy.

A common source of the agricultural adjustment problem is slow growth in the demand for food relative to the capacity to generate technological progress in agricultural production. In affluent industrial economies population growth is slow – typically slower than 1 per cent per year – and *per capita* food consumption also increases very slowly due to the low income elasticity of food demand. On the other hand, the rate of increase in domestic food supply is high because of the high rate of investment in agricultural research, development and extension. As the result a persistent tendency develops for the shift in domestic food demand to lag behind that of domestic supply.

When the excess food supply is faced with the inelastic demand schedule, farm income and the rates of return to resources used in agriculture decline unless the resources are transferred from agriculture to non-agriculture at a sufficiently rapid rate. In reality, the intersectoral transfer of resources (especially labour) is a slow process involving changes in generation. It is difficult to reallocate labour from the rural to the urban sector at a rate rapid enough to achieve income parity between farm and non-farm populations: this is what Schultz (1953) called the 'farm problem'.

In Japan, the agricultural adjustment problem is much more severe than other industrial countries because superimposed on the Schultzian farm problem is the adjustment need arising from changes in comparative advantage. This very severe adjustment problem may be called the 'agricultural problem of the *nouveaux riches*', because it is rooted in the fact that Japan rose (within only two decades) from a developing country to an industrial superpower with *per capita* income among the highest in the world.

1.2.2 Declining Importance of the Food Problem

Rapid growth in agricultural protection in Japan has been based on the organized political efforts of farm producers, aimed at shifting a part of the agricultural adjustment cost to the non-farm population. But protection policies would not have been instituted unless they were accepted by the non-farm population. In general, resistance to agricultural protectionism tends to be lost in the process of economic development.

A relative contraction of the agricultural sector in the total economy reduces the burden of agricultural protection *per capita* of the non-agricultural population. Consumers' resistance to agricultural protection is reduced as the Engel coefficient decreases corresponding to income rises, and hence the effect of rising food prices on the cost of living diminishes. Moreover, in the course of economic development an increasingly larger portion of consumers' food expenditure is allocated to marketing and processing services and a smaller portion is allocated to raw foodstuffs produced in the domestic agricultural sector.

With the decline in the effect of raw foodstuff prices on the cost of living, agricultural protection comes to have smaller effects on the wage rate, and hence is less strongly resisted by not only consumers but also by business employers. The loss of resistance from industrialists and traders to high food costs was especially large during the big spurt of industrial development after postwar reconstruction, because Japanese industries became increasingly more capital- and knowledge-intensive (and hence less dependent on low wages).

The declining resistance to agricultural protectionism may be considered a process of declining importance of the 'food problem' in the Ricardo–Schultz sense. Schultz (1953) identified the 'food problem' as an agricultural problem dominant in low-income, high food-drain economies, in contrast to the 'farm problem' dominant in high-income, low food-drain economies. In low-income countries population growth is fast, typically about 2.5 per cent per year. Also, *per capita* food consumption increases rapidly in the low-income countries if *per capita* income is raised, because food consumption is yet far from saturation. While the food demand increases rapidly in low-income developing economies, the shift in their food supply tends to lag behind, because both the human capital and the organization of research, development and extension to generate a new stream of agricultural technology are not well developed.

When excess food demand is faced with an inelastic supply schedule, food prices rise sharply – often to such an extent that it results in a major disruption by the urban poor. In the longer run, the increased cost of living pushes up the wage rates of urban workers, and hence raises the cost of industrial production and reduces international competitiveness of manufactures, especially at the stage of economic development in which comparative advantage lies in labour-intensive industries. Thus defined, the food problem is not a problem within the agricultural sector but a major problem of the national economy. Aggravation of this problem may well result in the arrest of economic development in its early stage, as feared by classical economists like David Ricardo at the time of the Industrial Revolution in England.

In Japan, too, the food problem had been dominant from the beginning of its modern economic growth since the Meiji Restoration (1868) until the postwar spurt of industrial development. During that period the agricultural adjustment problem had increasingly become serious, as comparative advantage had continued to turn against agriculture. However, so far as the economy had remained in the stage dominated by the food problem, agricultural protection policies had been resisted strongly by industrial and commercial interests, as well as by consumers. As the food problem lost its ground as the result of the spurt of postwar industrial development, domestic resistance against agricultural protectionism was lost and the agricultural lobby successfully raised protection to a level among the highest in the world. The high level of agricultural protection in Japan today is thus considered an equilibrium in the political market (see Appendix A, p. 129).

A major factor setting a limit on agricultural protection in Japan today is not domestic protests from the non-agricultural population, but foreign protests from Japan's trade partners. When domestic countervailing power has been lost in the course of economic development, the foreign protests have emerged as a substitute. Agricultural trade friction is thus so deeply rooted in the basic structure of political economy in Japan that it is extremely difficult to solve.

1.3 PLAN OF THE BOOK

The historical process of a shift from the food to the agricultural adjustment problem in the course of economic development in Japan since the early Meiji period is described in detail in Chapter 2, in order to give a proper perspective on today's problem. Between the stages of economic development dominated by the food and the agricultural adjustment problems an era is identified in which rural poverty became the dominant policy issue. Investigation is focused on the process in which the successive solution of the food and the poverty problems gave rise to the agricultural adjustment problem.

As the agricultural adjustment problem has become dominant, agricultural policy has been dominated by protectionism. In Chapter 3, the structure of agricultural protection in Japan is outlined by surveying various policy tools in relation to their goals and consequences. The mechanism of rice price support is analysed in detail as a typical example. It is shown that the protection policies moderate the problem of agricultural adjustment in the short run, but lead in the long run to a stalemate.

The long-run solution clearly lies in the adjustment of agrarian structure in a direction more consistent with the open trade system; this requires an expansion of operational farm sizes so that labour-saving technology by means of large-scale machinery can be efficiently utilized. Structural policies aimed at achieving this goal under the 1961 Agricultural Basic Law have not been successful because of the growing dominance of part-time farming. Factors underlying the failure of the 'Basic Law Agricultural Policy' are identified in Chapter 4.

More recently, however, a new prospect has been emerging concerning the possibility of structural adjustment. Chapter 5 analyzes the process in which scale economies have been created with a shift from small-scale mechanization based on hand tractors to large-scale mechanization based on riding tractors (as distinguished from power tillers, which are pushed as the operator walks along). Corresponding to this process – together with the relaxation of regulations on land tenure – the concentration of land in the hands of large farmers is now in progress.

Finally, Chapter 6 discusses the direction of agricultural policy that may be able to solve the conflicts of international harmony with national food security and the welfare of domestic farm producers. The discussions focus on political barriers to policy reorientation.

2 Changing Nature of the Agricultural Problem: A Historical Perspective

Within a century of modern economic growth since the Meiji Restoration, Japan has emerged from a low-income developing country to be one of the richest nations in the world. Meanwhile, the role of agriculture in the national economy has undergone a drastic change. The problem of agriculture *vis à vis* the rest of the economy has correspondingly experienced a major change. In this chapter, the changing nature of the agricultural problem in the process of economic development will be outlined, in order to give a historical perspective on the current situation.

Readers whose time is too limited to consider the historical background may wish to skip this chapter and turn directly to the analysis of the current situation in Chapter 3. Those who are interested in more detailed historical accounts of agricultural development in Japan, may wish first to study the volume by Hayami *et al.* (1975).

2.1 AGRICULTURE IN ECONOMIC DEVELOPMENT

Historical changes in the agricultural problem corresponding to changes in the relative position of agriculture in total economy will first be discussed, on the basis of the data in Table 2.1.

2.1.1 *Per capita* Income and Industrial Structure

As shown in Column (1) of Table 2.1, average *per capita* income in Japan in 1885 (the first year for which the estimate of national income is available) was 630 US dollars in 1980 prices; this was comparable to the level of developing countries in South East Asia today (e.g., 430 dollars in Indonesia, 670 dollars in Thailand and 690 dollars in the Philippines in 1980). Before the Second World War, Japan's *per capita* income had risen to 1600 dollars by 1935, a level comparable with present NICs (e.g., 1520 dollars in Korea and 2050 dollars in Brazil). After the spurt of postwar economic development, Japan joined the group of advanced industrial nations characterized by *per capita* income in the order of 10 000 dollars.

Rapidly rising *per capita* income was accompanied by major changes in industrial structure, as shown in Columns (2) and (3) in Table 2.1. Already before the Second World War, the share of the agricultural sector (including forestry and fisheries) in the total labour force decreased from more than 70 per cent to less than 50 per cent. After 1945 it declined further, to 10 per cent in 1980. Meanwhile, agriculture's share in total GDP declined from 45 per cent in 1885 to less than 20 per cent in 1935, and to 4 per cent in 1980.

A basic factor underlying the relative contraction in the agricultural sector was changes in the demand structure. The share of food in total consumption expenditure (the Engel coefficient) shown in Column (4) declined from more than 60 per cent in the early Meiji period to only about 30 per cent today. Meanwhile, the share of domestically-produced raw foodstuffs in total food consumption decreased from nearly 70 per cent to 20 per cent, reflecting an increasing share of consumers' food budget allocated to marketing and processing services (Columns (5), (6) and (7)).

The product of the Engel coefficient and the share of domestic foodstuffs measures the effect of increases in foodstuffs prices at the farm-gate on consumers' cost of living. Data in Columns (4) and (5) therefore imply that a hundred years ago a 10 per cent increase in the producer price of agricultural products would have resulted in about a 5 per cent increase in the cost of living in non-farm households, but today the same rate of increase in agricultural prices will result in only a 1 per cent increase. Obviously, the Japanese economy in its early stage of development was vulnerable to the 'food problem' in the sense that the rise in agricultural product prices and the resulting increases in the cost of living and the wage rate represented a serious impediment to industrialization and economic growth. Later, the menace of the food problem receded as *per capita* income rose in the course of economic development.

TABLE 2.1 Agriculture in Japan's economic development, 1885–1980

	Real GDP per capita	Share of agriculture		Engel coefficient	Shares in food consumption			Agriculture/ Industry labour productivity ratio	Farm/Non-farm household income ratio	Agriculture/ Manufacturing Terms of Trade
		labour force	GDP		Domestic foodstuffs	Imported food	Marketing and processing service			
	(1)[a]	(2)[a]	(3)[a]	(4)[a]	(5)[a]	(6)[b]	(7)[a]	(8)[a]	(9)[a]	(10) 1885=100
1885	630	73	45	64	66	1	33	75	76	100
1890	710	71	48	66	72	4	24	67	87	115
1900	880	68	39	62	53	4	43	49	52	102
1910	1000	65	32	61	47	5	48	37	47	98
1920	1260	54	30	62	47	7	46	50	48	99
1930	1350	50	18	53	30	9	61	31	32	104
1935	1620	47	18	50	36	12	52	24	38	136
1955	1850	39	21	52	55	8	37	55	77	163
1960	2690	32	13	43	47	5	48	39	68	169
1970	6920	17	7	34	34	7	59	25	91	304
1980	9890	10	4	31	23	8	69	17	115	347

Notes:
a US$ in 1980 prices.
b (%)

Column Notes and Sources

(1) Nominal GNP *per capita* in terms of yen deflated by the GNE deflator, based on K. Ohkawa and M. Shinohara (eds), *Patterns of Japanese Economic Development* (Yale University Press, 1979) (O-S Series), which are extrapolated to 1980 by the real *per capita* GDP series in the *National Economic Accounting* of the Japan Economic Planning Agency (JEPA) and linked to the *per capita* GDP in 1980 dollars in World Bank, *World Development Report 1982*.

(2) Agriculture includes forestry and fisheries. The O-S series until 1970, extrapolated to 1980 by Japan Prime Minister's Office, *Rodo Ryoku Chosa* (Labour Force Survey).

(3) NDP ratio for 1885–1935. The O-S series until 1970 and the JEPA series for 1975–80.

(4) The share of food consumption expenditure in total private consumption expenditure in current prices. The O-S series until 1970 and the JEPA series for 1975–80.

(5) The share of domestically supplied foodstuffs in total food consumption expenditure. The value of domestically supplied foodstuffs is calculated by subtracting from the value of domestic agricultural output those of intermediate agricultural products (seeds and feed), industrial crop output and foodstuffs exports. The output values from the O-S series and S. Yamada, 'Japan', paper presented at the Asian Productivity Organisation (APO) conference on *Agricultural Productivity Measurement and Analysis* (Tokyo, 1984). The export values from K. Ohkawa *et al.* (eds), *Long-Term Economic Statistics of Japan* (LTES) vol. 14 (Tokyo: Toyokeizai-shimposha, 1979); United Nations, *Yearbook of Trade Statistics*.

(6) The imported food values from LTES, vol. 14.

(7) 100 - (5) - (6).

(8) The ratio of real GDP per worker in agriculture, forestry and fishery to real GDP per worker in mining and manufacturing. The numbers of workers in fishery for 1880–1900 are assumed as 3 per cent (the 1906–10 average ratio) of the workers in agriculture and forestry. The numbers of workers in mining and manufacturing for 1880–1900 are assumed as 72 per cent (1906–10 average ratio) of the total labour force minus those of agriculture, forestry, fisheries, commerce and services. Data from the O-S series; the JEPA series; LTES, vol. 1; the *Labour Force Survey*.

(9) The ratio of average income per family member in farm households to that of urban worker households. The 1880–1935 data from T. Otsuki and N. Takamatsu, 'On the Measurement of Income Inequality in Prewar Japan', The International Development Centre of Japan (1982). The 1955–80 data calculated from Japan Ministry of Agriculture, Forestry and Fisheries, *Farm Household Economy Survey*; and Japan Prime Minister's Office, *Household Survey*.

(10) The ratio of the price index of agricultural products to the price index of manufacturing products from LTES, vol. 8, which is extrapolated by the producer price index of agricultural products of the JMAFF and the Bank of Japan's Wholesale Price Index of manufacturing commodities.

2.1.2 Agriculture–industry Relative Income

The relative contraction of the agricultural sector in the course of economic development is a universal phenomenon, mainly explained by the low income elasticity of demand for food. This process was augmented in Japan by a shift in comparative advantage from agriculture to industry corresponding to a rapid decline in real labour productivity in agriculture relative to that of industry (Column (8) in Table 2.1).

The rates of growth in agricultural output and productivity in Japan in the course of modern economic growth were not low when compared with the histories of other developed countries (Hayami and Ruttan, 1985). However, output and productivity in the manufacturing sector of Japan grew at exceptionally fast rates. The basic development strategy of Japan since the Meiji Restoration has been to catch up with western powers by importing their technologies, especially in modern industries. As a latecomer to industrialization, Japan was able to achieve a high rate of industrial productivity growth by exploiting the considerable potential available for technology borrowing. The persistent lag of labour productivity growth in agriculture behind manufacturing during the past 100 years in Japan was similar to that being experienced by NICs today (see Table 1.4, p. 11).

Until the Second World War, a decline in the agriculture–industry productivity ratio in real terms had been largely paralleled by a decrease in the ratio of average *per capita* income in farm households to that of non-farm households in nominal terms, reflecting a relative constancy in the domestic terms of trade between the agricultural and the manufacturing sectors (Columns (9) and (10) in Table 2.1). Such relations changed drastically after 1945. Despite the continued decline in the relative productivity of agriculture, the *per capita* farm income increased faster than the non-farm income – and, finally (in 1980), the former exceeded the latter by 15 per cent. The relative income position of farm households improved after 1945 because the domestic terms of trade improved greatly in favour of agriculture, in addition to rapid increases in off-farm income in farm households (Table 4.9 in Chapter 4, p. 92).

The improvement in the domestic terms of trade for agriculture in Japan after the Second World War was not a result of the free market mechanism, but of changes in industrial protection policies. As shown in Table 2.2, the terms of trade in the domestic market of

TABLE 2.2 Changes in agricultural and manufacturing commodity prices in Japan and the world

Commodity		*1960*	*1970*	*1980*
Japan:				
Agricultural product prices	(1)	100	195	432
Manufacturing product prices	(2)	100	109	211
Terms of trade	(1)/(2)	100	179	205
World:				
Unit value of agricultural exports	(3)	100	107	307
Unit value of manufacturing exports	(4)	100	115	340
Terms of trade	(3)/(4)	100	93	90
Japan/World price ratio:				
Agricultural products	(1)/(3)	100	182	141
Manufacturing products	(2)/(4)	100	95	62

Source: JMAFF, *Noson Bukka Chingin Chosa* (Statistics of Rural Prices and Wage Rates); The Bank of Japan, *Oroshiuri Bukka Shisu* (Wholesale Price Indexes); United Nations, *Statistical Yearbook of the United Nations*; FAO, *Production Yearbook* and *The State of Food and Agriculture*.

Japan (row (1)/row (2)) improved in favour of agriculture from 1960 to 1980, while it deteriorated in the world market (row (3)/row (4)). Such divergent trends reflect both the decreasing trade protection of manufacturing commodities and the increasing protection of agricultural commodities corresponding to the shift in Japan's comparative advantage from agricultural to manufacturing production. While the protection of the manufacturing sector was reduced as it left the 'infant-industry' stage, protection of the agricultural sector was raised in order to moderate the adjustment of the declining sector.

2.1.3 Three Agricultural Problems

It is clear that, until 1900 (or perhaps even until the First World War), the Japanese economy had been in a stage dominated by the 'food problem' and (since the 1960s) has entered a stage dominated by the 'agricultural adjustment problem'. Indeed, during the former period agricultural policies were oriented mainly towards increasing domestic production and a marketable surplus of food, whereas in the latter period policies have been geared towards supporting farm income by various protective means.

Between the eras of the 'food' and 'agricultural adjustment problems', it seems possible to identify an era in which public concern about agriculture was dominated by the 'poverty problem'. As already observed in Table 2.1, throughout the period before the Second World War the growth in labour productivity in agriculture lagged behind that of manufacturing but (unlike the period after 1945), agricultural protectionism was not strengthened to such an extent as to raise the domestic terms of trade appreciably in favour of agriculture. As a result, *per capita* farm income declined from about 80 per cent of non-farm income before 1900 to less than 40 per cent in the 1930s. The problem of rural poverty thus became a source of social instability and disruption during the interwar years, and social policies geared for relieving the poverty of peasants became a prominent feature of Japanese agriculture.

The interwar years may be considered a period in which Japanese economy reached the peak of the Kuznets inverted U-curve of income inequality. Why, then, was agricultural protection not strengthened at this time, unlike the recent period? The reason was the very strong opposition to agricultural protectionism. In this period, the pressure of rising food prices on the cost of living was still strong, especially among manual (blue-collar) workers. At the same time, the industries that had international competitive power were mainly the ones based on labour-intensive technologies and cheap labour. The increase in agricultural protection in those days to a level comparable with that of today would have endangered the process of industrialization and economic growth and hence been strongly resisted by the non-farm sectors.

During the interwar period the agricultural adjustment problem thus became serious as comparative advantage shifted away from agriculture owing to very rapid growth in industrial productivity, but the Japanese economy was not yet free from the menace of the food problem. Despite a growing rural–urban disparity, agricultural protection was kept under a certain limit. In these circumstances, the poverty problem emerged as the dominant agricultural problem.

2.2 OVERCOMING THE FOOD PROBLEM

When Japan's doors were opened to foreign countries shortly before the Meiji Restoration, the country was in real danger of colonilization by the western powers. The national slogan of the Meiji regime was to 'build a wealthy nation and strong army' (*Fukoku Kyohei*). To attain this goal it was considered necessary to 'develop industries and promote enterprises' (*Shokusan Kogyo*). Given industrial development as a prime national goal, agricultural policies in Meiji Japan were geared to securing sufficient domestic food supplies to prevent both a rise in the cost of living of urban workers and a serious drain on foreign exchange needed for the imports of industrial capital goods and technology. Policy was then focused on how to increase the marketable surplus of food products – especially rice, which was by far the most important wage good for urban workers.

2.2.1 Policies for Extracting a Marketable Surplus

The policy which contributed significantly to increasing the marketable surplus of rice in the early Meiji period was the Land Tax Revision (1873–6). During the Tokugawa period (which lasted some three hundred years prior to the Meiji Restoration), peasants who were given hereditary usufruct rights on farm land from feudal lords were, in principle, obliged to pay feudal tax in kind in proportion to harvested crop. The tax was assigned to each village, which was allotted by elected village officers (*Mura Yakunin*) to individual peasants. Later, the fixed land tax in kind became increasingly common because of its ease of collection. Meanwhile, the productivity of land increased and the gap between the tax and the rate of return to land widened. Through the illegal practice of land mortgaging and leasing, some of peasants became *de facto* landlords who collected rent from tenants and paid tax to feudal lords.

The Meiji government formally changed the feudal land tax to a modern tax in cash based on the assessed value of land. With this revision owner farmers and landlords were forced to market nearly one-quarter of rice output in order to pay the new land tax in cash.

The Land Tax Revision had another important consequence; it concentrated property titles in the hands of landlords. In the new regime, the modern property rights on land were granted to those who had previously held the traditional usufruct rights. Land tenancy became a legal institution. Because the new land tax was fixed in cash, small peasants were often unable to pay the tax in years of bad harvest or low rice prices. They were compelled to borrow money from wealthier farmers or landlords, and many of them lost their land through foreclosure. This process accelerated during the so-called 'Matsukata Deflation' in the mid-1880s, which was caused by the withdrawal of non-convertible paper currencies for the purpose of establishing the gold standard under direction of Finance Minister Masayoshi Matsukata. This deflation policy resulted in a drastic decline in agricultural product prices and raised the real value of agricultural debt. The area of land under tenancy was less than 30 per cent of total arable land area at the time of survey for the Land Tax Revision. It rose to 40 per cent in 1892, and to nearly 50 per cent by 1930 (Ouchi, 1960).

The growing concentration of land ownership, however, did not imply the polarization of peasantry into large estates and landless labourers. As shown in Table 2.3, the traditional agrarian structure in terms of a 'unimodal' distribution of small-scale family farms persisted (or was even strengthened) as the landlords stopped farming under their direct management and rented out their land to tenants in small parcels. In fact, this process had already begun in the eighteenth century with the progress of commercialization and the development of labour-intensive and land-saving technologies (Smith, 1959). From the late Tokugawa to the Meiji period, landlords took active leadership in the technology development and diffusion as a major means to increase land rent. It is considered that their supremacy in technological knowledge was an important factor underlying the concentration of farm land in their hands (Hayami *et al.*, 1975, pp. 52–4).

Before the Second World War, rent for paddy fields was paid in kind, roughly 50 per cent of the harvest. With the increasing share of land under tenancy the landlords – who had a much lower marginal propensity to consume rice – received an increased share of rice output. This in turn contributed to the increase in the surplus of marketable rice. Rice export in the early Meiji period was supported by the squeeze on farmers' incomes by heavy land tax and rent.

2.2.2 Policies for Increasing Agricultural Production

A more positive measure to increase the marketable surplus of food was to increase agricultural production. Meiji Japan inherited from the Tokugawa period a very unfavourable endowment of agricultural land resources relative to population and labour force, even when compared with the densely populated countries in East and South Asia. Given such initial conditions, how was Japan able to increase agricultural production? A detailed description of agricultural growth process in modern Japan is given elsewhere (Ogura, 1963; Hayami *et al.*, 1975). Here, only a brief sketch is given.

Year	No. of Farms (000)[a]							Average farm size
	Less than 0.5[b]	0.5–1	1–2	2–3	3–5	Larger than 5	Total	
1908	2016	1764	1055	348	163	62	5408	1.0
	(37.3)	(32.6)	(19.5)	(6.4)	(3.0)	(1.1)	(100.0)	
1910	2032	1789	1048	322	156	71	5417	1.0
	(37.5)	(33.0)	(19.3)	(5.9)	(2.9)	(1.3)	(100.0)	
1920	1935	1829	1133	341	154	92	5485	1.1
	(35.3)	(33.3)	(20.7)	(6.2)	(2.8)	(1.7)	(100.0)	
1930	1891	1892	1217	314	128	70	5511	1.1
	(34.3)	(34.3)	(22.1)	(5.7)	(2.3)	(1.3)	(100.0)	
1940	1796	1768	1322	309	119	76	5390	1.2
	(33.3)	(32.8)	(24.5)	(5.7)	(2.2)	(1.4)	(100.0)	
1950	2531	1973	1339	208	77	48	6176	1.0
	(41.0)	(32.0)	(21.7)	(3.4)	(1.2)	(0.8)	(100.0)	
1960	2320	1923	1430	233	91	60	6057	1.0
	(38.3)	(31.7)	(23.6)	(3.8)	(1.5)	(1.0)	(100.0)	
1970	2025	1614	1286	256	90	71	5342	1.1
	(38.0)	(30.2)	(24.1)	(4.8)	(1.7)	(1.3)	(100.0)	
1980	1938	1311	989	249	102	72	4661	1.2
	(41.6)	(28.1)	(21.2)	(5.3)	(2.2)	(1.5)	(100.0)	
1985	1870	1187	891	242	109	76	4376	1.2
	(42.7)	(27.1)	(20.4)	(5.5)	(2.5)	(1.7)	(100.0)	

Notes:
a Percentage distribution is shown in parentheses.
b All measurements in hectares.

Source: N. Kayo (ed.) *Kaitei Nihon Nogyo Kiso Tokei* (The Basic Statistics of Japanese Agriculture, revised edn) (Tokyo: Norin Tokei Kyokai, 1977); JMAFF, *Noringyo Census* (Census of Agriculture and Forestry) and *Kochi Oyobi Sakutsuke Menseki Tokei* (Statistics of Cultivated Land Areas and Areas Planted in Crops).

A century of agricultural growth in Japan since 1880 may be summarized as in Figure 2.1; the indexes of total agricultural output (Y), total input in agriculture (I) (which is the weighted sum of labour, land, capital and current inputs used in agriculture), and total productivity (Y/I) (which may be considered a proxy of technical change) are plotted in the upper graph; the indexes of labour productivity (Y/L), land productivity (Y/A) and land–labour ratio (A/L) in agriculture are plotted in the lower graph. By identity the following relations hold:

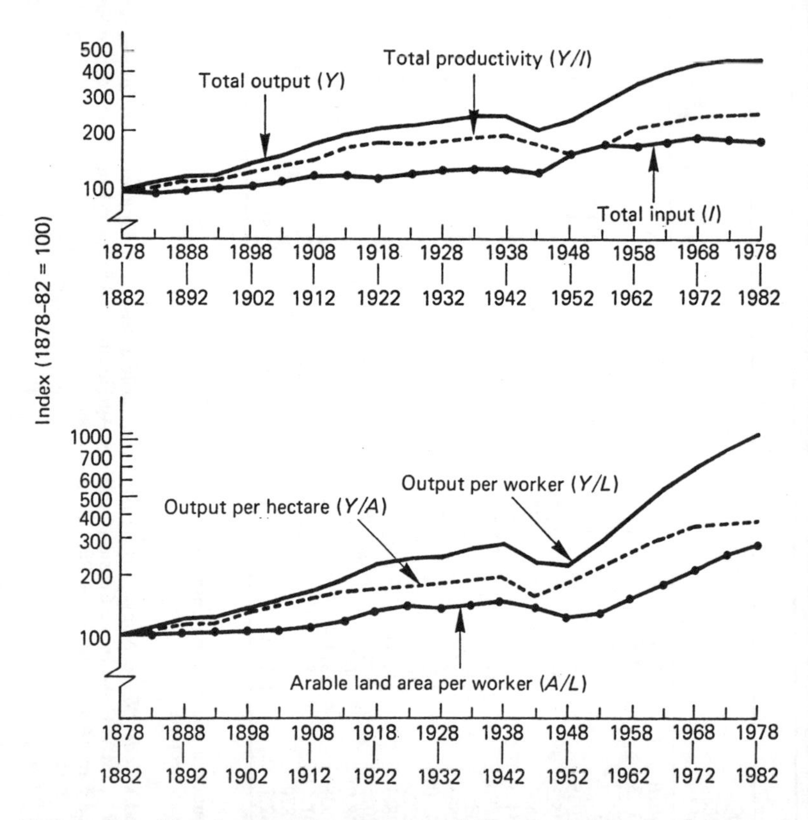

FIGURE 2.1 Trends in total output, total input, total productivity, and labour and land productivities in Japanese agriculture (stock terms, five-year averages, semi-log scale)

Source: Y. Hayami *et al.*, *A Century of Agricultural Growth in Japan* (University of Tokyo Press and University of Minnesota Press, 1975) with the 1973–82 data updated

$$Y \equiv I \times (Y/I) \text{ and } (Y/L) \equiv (Y/A) \times (A/L)$$

Overall, total output (Y) and labour productivity (Y/L) moved together. In their growth trends, three phases can be distinguished: (*a*) relatively fast growth up to the First World War, (b) relative stagnation in the interwar period, and (c) a spurt following the Second World War.

Critically important to the solution of the food problem was the fast growth in agricultural output and productivity in the initial phase. From 1880 to 1920, total output and labour productivity in agriculture increased in real terms at the annual average rates of 1.8 and 2.1 per cent respectively. Those rates were quite high compared with the historical records of Western Europe and the US (Hayami and Ruttan, 1985, Appendix B). The growth rate of agricultural output exceeded the population growth rate during this period (1.2 per cent per year). Therefore, even though *per capita* food consumption increased, a drain of foreign exchange from the import of foodstuffs was kept marginal. Instead, the agricultural sector acted as a major earner of foreign exchange through the export of silk and other agricultural commodities.

During the initial growth phase, acceleration in the growth rates can be observed. From the 1880–1900 to the 1900–20 period, the growth rate of total output increased from 1.6 to 2.0 per cent per year, and that of labour productivity from 1.6 to 2.6 per cent. This acceleration in agricultural growth was parallel with that of industrial growth. It was a major factor preventing the domestic terms of trade from turning against modern industries despite rapid increases in employment and income in the non-farm sector in the take-off period.

In Figure 2.1 it can be observed that the movements in total output (Y) are largely explained by those of total productivity (Y/I) rather than total input (I) and that the movements in labour productivity (Y/L) are explained largely by those of land productivity (Y/A) rather than the land–labour ratio (A/L). In fact, about 75 per cent of total output growth from 1880 to 1920 was accounted for by total productivity growth and 25 per cent by the growth in total input; about 65 per cent of labour productivity growth was accounted for by land productivity growth and 35 per cent by the growth in the land–labour ratio.

Such quantitative evidence indicates that, even in the initial growth phase, Japanese agriculture followed the pattern of modern economic growth depicted by Kuznets (1966) and Schultz (1953), in which a major portion of growth in output and output per worker resulted not so much from increases in conventional capital and land as from technological progress ('technological progress' is defined here as gains in output per unit of the conventional input). The evidence further suggests that the technological progress in Meiji Japan was mainly of the land-saving type, geared to raising yields per unit of scarce agricultural land. This hypothesis is also supported by the results of an econometric analysis (Kawagoe, Otsuka and Hayami, 1986). A key to the success in solving the food problem in the Meiji period (1886–1911) may be identified as the sustained progress in land-saving technology.

How, then, was Japan able to achieve significant progress in land-saving technology from the early phase of modernization?

Productivity growth in agriculture – the dominant sector of the economy – was required not only to meet the growing demand for food and foreign exchange, but was also essential to finance capital for industrialization and various modernization measures. Leaders of the Meiji government felt this need keenly, and policies for agriculture were geared to increasing output and productivity; they were commonly called *Kanno Seisaku* (Policies to Encourage Agricultural Production).

Initially, the Meiji government tried to develop agricultural production through the importation of western technology. During the 1870s large-scale machinery employed in England and the United States, and exotic plants such as olives and grapes for wine were imported and exhibited in experimental farms. Such early trials of direct 'technology borrowing' represented one example of the broad efforts of Meiji Japan to catch up with the western technology. Unlike the case in industry, however, this attempt was largely unsuccessful. The factor endowments and the scale of Japanese farming were simply incompatible with large-scale machinery of the Anglo-American type. In most cases, the efforts to transplant foreign plants and livestock also proved unsuccessful because of differences in ecological conditions.

The government perceived these failures, and redirected agricultural development strategy during the 1880s. The new strategy was to identify the best seed varieties and the most productive cultural practices used by farmers, improve them with the aid of new plant physiology and soil chemistry in the von Liebig tradition and propagate them nationwide. Unlike the earlier emphasis on exotic plants

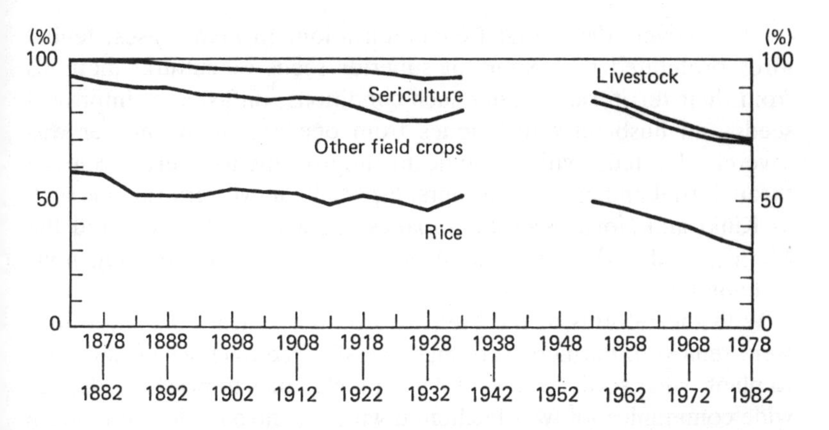

FIGURE 2.2 Composition of agricultural production in Japan by commodities (nominal composition, current prices, five-year averages)

Source: Y. Hayami *et al.*, *A Century of Agricultural Growth in Japan* (University of Tokyo Press and University of Minnesota Press, 1975) with the 1973–82 data updated

and animals, the new emphasis was placed on traditional crops, especially rice which accounted for about half of total agricultural output value during the Meiji period (see Figure 2.2). The government then employed veteran farmers (*rono*) as instructors in agricultural colleges and extension systems. In order to provide better information to the extension systems, agricultural experiment stations were set up for screening and tailoring veteran farmers' techniques by means of relatively simple comparative tests.

Such simple experiments provided a basis for the rapid growth of agricultural productivity during the latter years of the Meiji period because substantial indigenous technological potential could be further tested, developed, and refined at the new experimental stations. In addition, farmers with whom the research workers interacted effectively showed a strong propensity to innovate.

During the 300 years of the Tokugawa period preceding the Meiji Restoration farmers were subject to the strong constraints of feudalism. Personal behaviour and economic activity were highly structured within a hierarchical system of social organization. Farmers were bound to their land and were, in general, not allowed to leave their village. Neither were they free to choose what crops to plant or what varieties of seeds to sow. Barriers that divided the nation into feudal estates actively discouraged communication. In many cases, feudal lords prohibited the export of superior seeds or cultural methods from their territories. Under such conditions, diffusion of improved seeds and husbandry techniques from one region to another was severely limited, while significant improvements were achieved through trial and error of farmers, especially in western regions such as Kinki and North Kyushu. Japanese agriculture thus entered the Meiji period with a substantial backlog of unexploited indigenous technology.

With the reforms of the Meiji Restoration, such feudal restraints were removed. Farmers were made legally free to choose what crops to plant, what seeds to sow, and what techniques to practise. Nationwide communication was facilitated with the introduction of modern postal service and railroads. The cost of information diffusion concerning new technology was greatly reduced. The land tax reform, which granted a fee-simple title to the farmers and transformed a feudal sharecrop tax to a fixed-rate cash tax, increased farmers' incentive to innovate. Farmers, especially of the *gono* class (landlords who personally farmed part of their holdings), vigorously responded to such new opportunities. They voluntarily formed agricultural societies called *nodankai* (agricultural discussion society) and *hinshukokankai* (seed exchange society) and searched for profitable techniques.

The veteran farmer techniques were based on experiences in the specific localities where they originated. They tended to be location-specific and to require modification when transferred to other localities. Simple comparative tests effectively screened the *rono* techniques and varieties, thereby reducing the risk to farmers adopting the new technology. Slight modifications or adaptations of indigenous techniques on the basis of experimental tests often gave them universal applicability.

The technology developed in this process was, by necessity, motivated to increase land productivity or was biased toward saving land, which represented the major constraint on agricultural production in Japan. However, it required an increase in the application of the input that substitutes for land – that is, fertilizer. Agricultural growth based on the improvement in indigenous technology was supported by an improvement in the supply of fertilizers, which depended on active entrepreneurship in farm supply industries.

A prerequisite for such development in the seed-fertilizer technology was an irrigation and drainage system that adequately controlled the supply of water. In this respect, Meiji Japan was favoured by the inheritance of a relatively well developed land infrastructure from the feudal period. Improvements in irrigation system had been the major source of productivity growth in wet rice culture in the feudal period. It had been the prime responsibility of feudal rulers to control rivers and major irrigation systems. Village communities had taken the responsibility of maintaining and improving irrigation infrastructure at the village level by mobilizing communal labour, while the choice of cultural practices and varieties in rice production itself had been left largely to the decision of individual producers within the limit of communal irrigation management. The irrigation development had been promoted continuously during the long peaceful Tokugawa period, in which land had become increasingly scarce corresponding to the growth in population. The increasing need of villagers to cooperate for irrigation construction and management was a major factor underlying the 'tightly-structured' social system in Japanese villages, in contrast with the 'loosely-structured' system in some of land-abundant economies in South East Asia such as Thailand (Hayami and Kikuchi, 1981, pp. 20–1). By the late Tokugawa period, rice cultivation in dry upland condition had become negligible. In 1883 (the first year for which official statistics were published), the share of upland rice in the total area planted in rice in Japan was only about 1 per cent, and its share of rice output was about 0.5 per cent.

By the beginning of the Meiji era, nearly 100 per cent of paddy field area in Japan had already been irrigated. However, water supply had often been insufficient and drainage facilities lacking in many fields. In order to facilitate the land-improvement projects, the Meiji government enacted the Arable Land Replotment Law in 1899 (revised in 1905 and 1909). By the law: (a) participation in the project was made compulsory upon the consent of more than two-thirds of the landowners owning more than two-thirds of the arable land area in the district concerned, and (b) a legal-person status was given to the associations of land-improvement projects, so that they could receive credit. The government also established the Japan Hypothec Bank in 1897, which aimed to advance long-term credit for land-infrastructure investments. As a supplement to the Hypothec Bank, the Banks of Agriculture and Industry were established, one for each prefecture (1880–1900). Later, the government advanced low-interest credit from funds mobilized from the postal savings for land-improvement projects through the Hypothec Bank and the prefectural Banks of Agriculture and Industry. Both central and local governments promoted the land-improvement project by giving subsidies through agricultural associations.

The nationwide diffusion of the seed-fertilizer technology progressed on the basis of improvement in land infrastructure. It originated in advanced areas in western Japan, where the population density was higher and the irrigation infrastructure better established. Later, beginning in the first decade of the twentieth century, this process was transmitted to the relatively backward eastern part of Japan.

2.2.3 From Tariff to Imperial Food Self-sufficiency

The government policies of developing agricultural production, when supported by the high propensity of landlords to innovate, were successful in preventing the food problem from becoming a major impediment to industrialization and economic growth during the Meiji period. The early Meiji policies to encourage agricultural production were formed when the interests of landlords and industrial capitalists coincided. In the early stage of industrialization (in which agriculture's comparative advantage was still high, and demand for the produce of domestic agriculture was highly elastic with respect to both income and price) the landlords were able to increase their income by increasing agricultural output via increasing the productivity of land. The capitalists also wanted larger domestic agricultural production as a source of cheap food for their workers. Such coincidence of interests, however, was short-lived.

Based on the success of agricultural development, modern industries grew rapidly and comparative advantage shifted away from agriculture. The interest of the landlord class thus began to shift from an increase in agricultural production to political lobbying for protection. The situation was such that 'the issue voiced most loudly in the general assemblies of the Imperial Agricultural Society was not the increase in productivity but the increase in tariffs' (Tobata, 1956, p. 597). This shift in the interest of the landlord class reflected a shift in their role from an innovator in agricultural production to a mere receiver of land rent and a siphon of agricultural surplus to non-agricultural sectors – the so-called 'transformation from entrepreneurial landlords to parasitic landlords (Ouchi, 1960, pp. 85–96).

Voices for tariff protection of rice in terms of foreign exchange and national security considerations began to be raised when Japan became a net importer of rice in the 1890s. But voices favouring securing cheap rice for cheap labour were equally strong. It was in the first year of the Russo-Japanese War (1904–05) that the 15 per cent *ad valorem* tariff was first imposed on imported rice.

This tariff was levied to increase the government revenue for financing the war, and it was to be terminated at the war's end. Yet the landed interests lobbied to preserve this tariff, and succeeded in making it permanent in 1906 in the form of a specific duty of 0.64 yen per 60 kilograms. Subsequently, the rice tariff became a major issue of public controversy (similar to those over the British Corn Laws and the German grain tariffs). Like Robert Malthus and David Ricardo on the Corn Laws and Adolf Wagner and Lujo Brentano on the German grain tariff, Jikei Yokoi in the University of Tokyo and Tokuzo Fukuda in the Tokyo College of Commerce (Hitotsubashi University) represented the two camps. Yokoi, the foremost leader of agricultural fundamentalism, argued for the tariff on the basis of national security, including the preservation of agriculture as the source of strong soldiers and considerations of the balance of payments and the balanced growth of agriculture and industry. Fukuda retaliated on the basis of the economic doctrine of the Manchester school which favoured free trade and industrial growth. The controversy continued in the arena of the National Diet. The Imperial Agricultural Society (representing the landed interests) and the Tokyo Chamber of Commerce (representing the manufacturers and traders of export goods) lobbied fiercely for opposite ends (Mochida, 1954).

In 1913, the rice tariff controversy in the Diet ended with the imposition of a specific duty of one yen per 60 kilograms, which could be lowered by 0.4 yen by government order. An important qualification was that the import of rice from overseas territories (Taiwan and Korea) was made free of duty. This decision pointed to the possibility of solving two conflicting policy goals – (a) self-sufficiency of rice and (b) supply of cheap rice for urban workers – by expanding the source of rice supply from Japan to its overseas colonies. This policy of imperial self-sufficiency was not deliberately adopted before the *Kome Sodo* (Rice Riot) in 1918. Rather, 'it had been considered that the development of rice production in those overseas territories should be suppressed since it was to foster competition against Japanese agriculture' (Tobata, 1956, p. 597).

Increases in rice yield and production, which had kept rice imports from rising, began to slow down in the mid-1910s. The technological potential which existed in the early Meiji period was now being exhausted as it became perfected and propagated (Hayami and Yamada, 1968). The agricultural experimental stations in their early days contributed to agricultural productivity growth by exploiting indigenous potential rather than by supplying new potential. The national experimental stations gradually came to conduct more basic research, including original crop-breeding projects at the Kinai Branch by cross-breeding (1904) and at the Rikuu Branch by pure line selection (1905). Results of major practical significance lagged, however, for more than two decades.

The exploitation of indigenous potential and the lack of new potential in scientific research, when confronted with the expansion of demand due to the First World War, resulted in a serious rice shortage and a high price for rice. These naturally caused disruptions in the urban areas and culminated in the Rice Riot of 1918, which swept over all the major cities in Japan.

Faced with a choice between high rice prices, high cost of living, and high wages on the one hand, and a drain on foreign exchange by large-scale rice imports on the other, Japan organized the imperial self-sufficiency programme. Under the programme *Sanmai Zoshoku Keikaku* (Rice Production Development Programme), the Japanese government invested in irrigation and water control and in research and extension to develop and diffuse high-yielding Japanese rice varieties adaptable to the ecologies of Korea and Taiwan. The success of this effort created a tremendous surplus which flooded the Japanese market. Within 20 years (from 1915–35), net imports of rice from Korea to Japan rose from 170 to 1212 thousand metric tons per year, and net imports from Taiwan rose from 113 to 705 thousand metric tons. Because of this inflow of colonial rice, the net import of rice rose from 5 to 20 per cent of domestic production.

With the success of the imperial food self-sufficiency policy, Japan was again able to overcome the food problem. But, this success paved the way for the poverty problem in the interwar period.

2.3 ERA OF THE POVERTY PROBLEM

After the First World War, the capital intensity of Japanese manufacturing increased markedly, corresponding to a shift in its centre of gravity from light to heavy industries, with the result that the productivity gap betweeen the agricultural and the manufacturing sectors widened further. With the concentration of capital in the hands of large-scale industries, the so-called 'differential structure' was created, in which both the productivity and the wage rates of workers in the large-scale industries were significantly higher than those of small- and medium-scale industries (Ohkawa, 1972; Minami, 1986, Chapter 9). Workers in the small-scale industries, together with those engaged in petty trades and other unskilled services, formed a reserve army of labour for the large-scale industries. A major source of labour supply for this reserve was the agricultural sector. In this economic structure, rural poverty was an anchor which prevented wage rates in the urban organized sector from rising. How to maintain the income and the living of small peasants without causing undue strain on social and political stability became the major problem of agricultural policies. In that sense, the 'agricultural problem' in the interwar period was dominated by the 'poverty problem.'

2.3.1 Policies in the Agricultural Crisis

The economic climate in the interwar period was generally unfavourable to agriculture. Demand for Japanese manufacturing products contracted with the recovery of Europe from the disruption of the First World War. The income (and the employment) in the urban sector correspondingly declined. With the increase in capital intensity in the manufacturing sector, the wage income of industrial workers also increased slower than industrial output. Those factors contributed to stagnation in the demand for food.

The success of the government programme to develop Korea and Taiwan as major suppliers of rice to Japan was a mixed blessing. Large-scale imports of rice, characterized by a relatively inelastic demand schedule, lowered the price and discouraged the production of rice in Japan. It was estimated that competition from the colonies lowered the domestic rice price in Japan by as much as 40 per cent in the mid-1930s (Hayami and Ruttan, 1970).

Already during the 1920s competition from colonial rice producers, together with the deflationary policy of the government to return to the gold standard at prewar parity, depressed agricultural prices and income. Finally, the world depression hit Japan, resulting in a serious agricultural crisis, and the government was compelled to support rice prices.

When the price of rice began to fall in the 1920s the Imperial Agricultural Society pressed the government to adopt a rice-control programme, the so-called Ever-Normal Granary Plan. This demand brought about the Rice Law of 1921, which empowered the government to adjust the rice supply in the market by: (a) operating the purchase, sale, storage, and processing of rice within the financial limit of 200 million yen, and (b) reducing or increasing the import duty and restricting imports from foreign countries.

In response to the rapid decline in rice prices in the late 1920s and 1930s, the Rice Law was amended in 1925, 1931, and 1932, raising the financial limit finally to 480 million yen. In 1933, when a bumper crop caused a phenomenal surplus of rice, the Rice Law was replaced by the Rice Control Law, which authorized the government to buy and sell unlimited quantities of rice at the floor and ceiling prices. The government rice-control operation was extended to colonial rice.

In addition to the farm price support, the government developed various programmes to mitigate the agricultural crisis. These included: (a) government spending on construction of physical infrastructure in rural areas in order to provide wage-earning opportunities; (b) release of low-interest loan from government to farmers heavily in debt from private money lenders; and (c) organization of economic recovery movements for villages that promoted self-sufficiency both in production inputs and in consumption goods, thereby reducing the cash expenditures of farm households. Those programmes were promoted by government subsidies channelled through agricultural associations and cooperatives.

In spite of all those efforts, the level of income and the living standard of farm people did not appreciably improve. The tenant farmers, who were especially hard hit, demanded rent reductions. By this period, the paternalistic relations between landlords and tenants which characterized the agrarian society of Meiji Japan had already faded as landlords had been transformed into a parasitic class. The tenant-union movements for rent reduction and the establishment of stronger tenancy rights gained momentum, and landlords retaliated, sometimes by foreclosing tenanted land. Subsequently, tenancy disputes became a common feature in rural areas in Japan. To cope with the situation, the government tried to undertake land reform programmes. But, barred by the strong political power of landlords, little was accomplished before the Second World War.

2.3.2 A Limit to Protection

The agricultural policies during the period between the two world wars were called the 'agricultural-cum-social policies' (Ouchi, 1960).

Unlike policies in the Meiji period that were oriented towards increasing production and a marketable surplus, the interwar policies were motivated towards protecting 'poor peasants' so that they would remain as an anchor to social stability as well as a reserve of cheap labour for the urban sector.

The shift in the policy orientation from exploiting agriculture for the sake of industrial development to subsidizing agriculture for the alleviation of rural poverty is best illustrated by changes in the sectoral allocations of tax burden and subsidies shown in Tables 2.4 and 2.5. In the early Meiji period (before 1900) the agricultural sector shouldered about 90 per cent of the total direct tax burden, which amounted to about 15 per cent of agricultural income. Direct taxes were only 2 per cent in the non-agricultural sector. During this period, subsidies from the national government were allocated exclusively to the non-agricultural sector; the tax burden of the agricultural sector was reduced relative to that of the non-agricultural sector and the share of agriculture subsidies thus rose rapidly. In the 1930s, when the problem of rural poverty became most serious under the influence of the world depression, the ratio of subsidy to income in the agricultural sector became higher than that in the non-agricultural sector.

The greater allocation of government subsidies to agriculture was partly a result of political lobbying by landlords. It was also a result of decreasing resistance by industrial and commercial interests to agricultural subsidies. In the interwar period, modern industries in Japan reached a self-sustaining stage, no longer so dependent on government supports as they had been in the Meiji period. They still needed cheap food for cheap labour, and on this score the interests of industrialists were in conflict with those of landlords. However, their interests coincided in the need to prevent peasants from becoming political allies to socialism and the labour-union movements that began to gain momentum in this period. Industrialists found it necessary to concede to the demand of landlords to some extent.

The coalition between the landed class and the industrial and commercial class was not unique to the history of modern Japan. In nineteenth-century Britain, industrial and commercial interests won the political battle against the landowners for the repeal of the Corn Laws in 1846. However, other major European powers that followed Britain in industrialization did not take the British route. For example, Germany (which was a traditional exporter of grain to Britain) moved from free trade to protectionism in the late nineteenth century as comparative advantage shifted away from agriculture as the result

TABLE 2.4 Changes in direct tax burdens on the agricultural and the non-agricultural sectors in Japan

	Direct tax burden[a]				*Direct tax rate*[b]	
	Agriculture		*Non-agriculture*		*Agriculture*	*Non-agriculture*
	(million yen)	*(%)*	*(million yen)*	*(%)*	*(%)*	*(%)*
1878–82	63.6	91	6.3	9		
88–92	58.5	86	9.8	14	14.9	2.0
98–02	99.1	74	35.4	26	11.7	2.7
1908–12	153.4	54	132.2	46	11.2	5.5
18–22	295.7	41	431.1	59	7.5	4.8
28–32	205.5	33	421.3	67	8.1	3.8
33–7	197.3	26	559.2	74	6.5	4.0

Notes:
a Includes both national tax and local rates.
b Ratio of direct burden to sectoral NDP.
Source: Tax data from S. Tobata and K. Ohkawa (eds), *Nihon no Keizai to Nogyo* (Economy and Agriculture in Japan), vol. 1 (Iwanami, 1956). Sectoral NDP data from K. Ohkawa and M. Shinohara, *Patterns of Japanese Development* (Yale University Press, 1979).

TABLE 2.5 Changes in the allocation of government subsidies to the agricultural and the non-agricultural sectors

	Subsidy receipt[a]				*Subsidy rate*[b]	
	Agriculture		*Non-agriculture*		*Agriculture*	*Non-agriculture*
	(million yen)	*(%)*	*(million yen)*	*(%)*	*(%)*	*(%)*
1881	0	0	0.7	100		
1891	0	0	2.5	100	0	0.49
1901	0.4	2	18.7	98	0.05	1.41
11	0.3	1	27.8	99	0.02	1.09
21	0.6	1	51.8	99	0.02	0.55
31	21.4	17	101.5	83	1.17	1.11
34	28.3	28	71.0	72	1.14	0.58

Notes:
a National government subsidies.
b Subsidy receipt divided by sectoral NDP.
Source: Tax data from S. Tobata and K. Ohkawa (eds), *Nihon no Keizai to Nogyo* (Economy and Agriculture in Japan), vol. 1 (Iwanami, 1956). Sectoral NDP data from K. Ohkawa and M. Shinohara, *Patterns of Japanese Development* (Yale University Press, 1979).

of successful industrialization and the inflow of cheap grains from the new continents. The tariff on grain imports introduced in Germany by Bismarck in 1879 was the result of a united campaign by the Junkers (large estate owners) in eastern Germany and the iron and steel industrialists in western Germany, who formed a 'solidarity bloc' which demanded protection against imports of both agricultural and industrial products.

Unlike Britain, which had established itself as the 'workshop of the world' by the beginning of the nineteenth century, comparative advantage in manufacturing was less certain in Germany, and industrialists found it advantageous to seek protection for their products, even at the expense of high food prices. In addition, the rapid growth of the Social Democratic Party (a labour party based initially on orthodox Marxist doctrine) was perceived as a common menace by the Junkers and the industrialists (Gerschenkron, 1943). This German experience was largely repeated in France and Italy (Kindleberger, 1951).

Growing agricultural protectionism in Japan during the interwar period can thus be considered an experience shared in common by late-comer countries to industrialization at the stage when they successfully completed a 'take-off'. However, agricultural protection in the interwar period had been limited mainly to the relief of rural poverty and never raised to a level sufficiently high to counteract the widening rural–urban income disparity. This must be evident from the estimates of nominal rates of protection on rice in Japan before the Second World War, which are presented in Table 2.6; the nominal protection rates were calculated by comparing the wholesale price and the import price of the indica rice imported from South East Asia, because there was a considerable price difference between the Japonica and the indica rice in the Japanese market. As measured in Column (4) in Table 2.6, the protection rate rose from the beginning of this century but (even in the agricultural crisis period of the 1930s) it remained below 50 per cent, which was incomparably lower than today's level (see Table 1.3, p. 9).

There was a limit to growth in agricultural protection before the Second World War, not because demand for protection was weak but because the food problem was still serious. As evident from the experience of the Rice Riot in 1918, the level of income of urban workers before the Second World War was such that the major increase in food prices could result in disruption. The international competitive power of the manufacturing sector was still dependent to

TABLE 2.6 Estimates of nominal rates of protection on rice in Japan before the Second World War

	Milled rice price (yen/ton)			*Nominal rate of protection*
	Wholesale price		*Cif price of South East Asian rice*	
	Japanese rice	*South East Asian rice*		$(4) = \frac{(2)-(3)}{(3)} \times 100$
	(1)	*(2)*	*(3)*	
1903–7	101	82	75	9
1908–12	118	93	77	21
1913–17	124	103	81	27
1918–22	286	188	165	14
1923–7	280	163	147	11
1928–32	187	144	114	26
1933–7	204	162	112	45

Source: K. Anderson and Y. Hayami, *The Political Economy of Agricultural Protection: East Asia in International Perspective* (Allen & Unwin, 1986).

a large extent on cheap labour, so that a major increase in the wage rate resulting from large increases in food prices could not be tolerated.

In such circumstances, with whatever powerful lobbying the landlords were able to organize, it was not politically possible to raise the level of agricultural protection sufficiently high to reduce the rural–urban income disparity. The 'agricultural-cum-social policies' were unable to prevent the agricultural crisis in the 1930s from turning the rural sector into a nursery of Fascism, leading to military intervention in China, and finally to the Pacific War.

2.4 TOWARDS THE DOMINANCE OF THE AGRICULTURAL ADJUSTMENT PROBLEM

During and immediately after the Second World War, the Japanese economy returned to a stage which was dominated by the 'food problem'. But, within two decades after 1945, Japan had jumped to a new stage, in which agricultural adjustment was the predominant agricultural problem. This process will be summarized below.

2.4.1 Food Policy During War and Recovery

When Japan waged the war against China (and later against the Allied Powers), it was found that the Imperial food self-sufficiency in peace was no guarantee of food security in crisis. After the 'China Incident' in 1937, shortages of labour and material inputs such as fertilizers were keenly felt, and a decline in domestic food supply became evident. Government reserves of rice rapidly decreased, and were exhausted in 1939 with the severe drought which hit western Japan and Korea.

In the progress of the war the government was forced to take direct control of rice distribution and began with the Rice Distribution Control Act in 1939. Increasing numbers of food items were added to the list of direct control and rationing. Finally, the Food Control Act, proclaimed in the second year of the Pacific War in 1942, put nearly all items of food under the strict control of the government.

The food problem became an especially critical issue in Japan during the recovery from war devastation. A compulsory delivery of rice from producers at a price far below market equilibrium had to be enforced to maintain the subsistence of a majority of the urban population suffering acute food shortage and hyperinflation. The consumer price for the government ration of rice was set even lower than producer and import prices.

In 1946, by the direction of the General Headquarters (GHQ) of the US occupation forces, the government introduced the parity price formula for the determination of the producer price of rice. Theoretically, this formula guaranteed the same terms of trade of rice with the commodities that farmers had bought in the base years (1934–6). However, the commodity prices taken in the calculation of the parity index were the official prices of government rationing. Since farmers had to rely on black markets to a large extent for the purchase of both production and consumption goods, the parity price did not really guarantee 'parity'. To secure the delivery of rice at below-equilibrium prices, the Food Emergency Measure Act was promulgated in 1946, empowering the government to expropriate undelivered rice. Several incentive schemes were also designed, giving bonuses to producers who completed delivery.

The government also made substantial investments in reclamation, irrigation, drainage, agricultural research and technical extension; increases in food production were considered a necessary condition in the design of industrial recovery. In the programme for the rehabilitation of industry, called *Keisha Seisanhoshiki* (Differential Production Scheme) that began in 1946, the government fund was first allocated to coal mining. Increased deliveries of coal were allocated to fertilizer, iron, and steel industries; increased deliveries of food from fertilizers together with those of iron and steel were returned to coal mining to expand the cycle of reproduction.

As industry was rehabilitated on the basis of compulsory deliveries of cheap food, agricultural production recovered with the increased supply of industrial inputs to agriculture. Government controls on food commodities were lifted one by one: potatoes in 1949, wheat in 1952, etc. It was planned to lift the direct control on rice in April 1952, but this plan was withdrawn because of the bleak prospects for food supply during the Korean War.

2.4.2 Postwar Institutional Reforms

Immediately after 1945, reforms for 'democratizing' agrarian communities were promoted under the direction of GHQ. Of special significance were the land reform and the reorganization of the agricultural cooperative associations.

The land reform was carried out in the 1946–50 period in accordance with the strong recommendations of the occupation authorities (Hewes, 1950; Dore, 1958; Ogura, 1963). The position of landlords had already been seriously undermined during the war. The urgent need to increase agricultural production by increasing production incentives to the cultivators had overcome the opposition of the landlords to strengthening the rights of tenants (the Farmland Adjustment Law 1938), to controlling rents (the Rent Control Order 1939), and to controlling land prices (the Farmland Price Control Order 1941). By the land reform laws of 1946 (the Revision of the Farmland Adjustment Law and the Owner-Farmer Establishment Special Measure Law), the government was authorized to enforce the purchase of all farmlands owned by absentee landlords as well as the land holdings of resident landlords exceeding 1 hectare (4 hectares in Hokkaido), which should be sold to tenants within two years after the proclamation of the law. To execute the land transfers, an Agricultural Land Commission was established in each village, consisting of three representatives from the landlords, two from the owner-farmers, and five from the tenant farmers.

The land prices paid to the landlords were determined as 40 times the annual rent in the case of lowland paddy fields and 48 times in the case of upland fields. In this formula the rents in kind were evaluated by the commodity prices of November 1945. Consequently (in the process of rapid inflation from 1945 to 1949) the real burden of tenant farmers in procuring land was reduced to a negligible level.

For the four years from 1947 to 1950 the government purchased 1.7 million hectares of farmland from landlords and transferred 1.9 million hectares, including state-owned land, to tenant farmers, which amounted to about 80 per cent of the land under tenancy before the land reform. As a result, the ratio of farmland under tenancy declined from 45 per cent in 1945 to 9 per cent in 1955. Further, for the remaining land under tenancy, the right of tenants was strengthened and the rent was controlled at a very low level by the Agricultural Land Law (1952). This Law also imposed a limit on landholding to 3 hectares (12 hectares in Hokkaido) in order to prevent the revival of landlordism.

The success of the drastic land reform in Japan was, to a large extent, based on the power of the occupation forces. Equally critical was the backlog of knowledge and experience of the land-tenure system, accumulated by the Japanese government through the trials of settling tenancy disputes during the interwar period. Also important were the various measures of controlling land tenures that had developed during the war with the effect of undermining the position of the landlords.

It must be recognized, however, that, although the land reform resulted in a considerable change in the distribution of landownership, the distribution of units of cultivation experienced no basic change. The traditional agrarian structure of Japan in terms of a 'unimodal' distribution of small-scale family farms with the average size of about 1 hectare remained essentially unchanged, irrespective of the rise and the fall of landlordism (see Table 2.2, p. 23).

There is no doubt that the land reform promoted more equal assets and income distributions among farmers, and hence contributed to the social stability of the rural sector. However, the reform did not induce changes in the basic direction of technological developments because, as before, the small-scale family farms remained the basic unit of agricultural production. Although land reform contributed to an increase in the level of living and consumption, its contributions to capital formation and productivity growth in agriculture are not significant in terms of quantitative analysis (Kawano, 1969).

Another reform, which also had a major impact on the agricultural economy and rural society, was the reorganization of agricultural cooperative associations. During the war, the agricultural associations and the agricultural cooperatives were integrated into a semi-governmental organization called *Nogyokai* (Agricultural Society) that was designed to share the responsibility of controlling and mobilizing village economies for war purposes. This organization was dissolved in 1947 by the direction of the GHQ. All the economic functions of the *Nogyokai*, including marketing and credit, were transferred to the agricultural cooperative associations reestablished by the Agricultural Cooperative Law in 1947.

The agricultural cooperative associations – commonly called '*NOKYO*', an abbreviation of *Nogyo Kyodo Kumiai* – inherited the nationwide organizations from the *Nogyokai*. 'General cooperatives' (*Sogo NOKYO*) (today numbering more than 4000) engage in marketing of all farm products and inputs together with credit and insurance business, each operating in the regional franchise of a village or a town. Those village cooperatives are organized into prefectural and national federations. The national federations at the top of the pyramid include the National Federation of Agricultural Cooperatives (*ZENNOH*) for marketing; the Central Bank of Agriculture and Forestry (*CHUKIN*) for credit; the National Insurance Agricultural Cooperative Federation (*ZENKYOREN*) for life and casualty insurance; and the Central Union of Agricultural Cooperatives (*ZENCHU*) for political lobbying. The organization of the *Sogo NOKYO* forms a formidable political bloc with some 300 000 employees and 5 million members (with additional 2 million associate members consisted mainly of non-farm residents in villages). Besides the network of general cooperatives, 'specialized cooperatives' (*Senmon NOKYO*) (numbering nearly 2000) are organized by the producers of specific products, such as the citrus-grower cooperatives and the dairy-producer cooperatives, with their own federations.

The development of the cooperative associations, especially *Sogo NOKYO*, was facilitated by pervasive government control on agricultural products and inputs for a few years immediately after the Second World War. The cooperatives used to monopolize the delivery of food products – above all, rice – to the Food Agency of the Ministry of Agriculture and Forestry, and the distribution of government ration of fertilizers and other inputs. Even today, probably more than 70 per cent of the rice and fertilizers is marketed through the agricultural cooperatives, although the exact volume of rice marketed through the illegal private channels cannot be identified. Their monopolistic position has also been strengthened by consignment of semi-governmental functions to them, such as channelling of low-interest institutional loans.

The basic feature of Japanese agriculture today in terms of a large number of small-scale owner-farmers, organized by *NOKYO*, was thus framed through the postwar reforms.

2.4.3 Emerging Need of Agricultural Adjustment

When Japan recovered from the devastation of the war and set off on its 'miraculous' economic growth after the mid-1950s, agriculture began to face serious adjustment problems.

The rate of growth in agricultural productivity, which was rapid by international standards, was not rapid enough to keep up with the growth in the industrial sector. Intersectoral terms of trade did not improve for agriculture during the 1950s, since the end of the Korean War – partly because of the pressure of surplus agricultural commodities in the US and other exporting countries, and partly because the domestic demand for major staple cereals (especially rice) approached saturation point after the bumper crop of 1955. In consequence, the levels of income and living standards of farm households lagged behind those of urban households during the 1950s.

In such a situation, the major goal of agricultural policy shifted from an increase in the production of food staples to a reduction in the rural–urban income gap. To attain this goal, the New Village Construction Programme was initiated in 1956. This programme was aimed at improving the rural economy by providing a package of economic and social infrastructure to the villages that prepared their own village-development plans. Each village was encouraged to design and promote its own agricultural development plan in order to increase the production of commodities having high-income elasticities, such as livestock, fruits, and vegetables. Long-term credits were advanced from the Agriculture, Forestry and Fisheries Credit Corporation, funded by the government, for the purchase of livestock, pasture improvement, and so on.

The need for assisting agricultural adjustments increased in the 1960s as the rural–urban income gap progressively widened and the outmigration of agricultural labour accelerated. These difficulties led to the enactment of the Agricultural Basic Law, a national charter for agriculture, in 1961 (Ogura, 1963, pp. 286–96). This Law declared that it was the government's responsibility to raise agricultural productivity and thereby close the gap in income and welfare between farm and non-farm people. Among the measures identified as necessary for this purpose were encouragements to expand selectively the production of agricultural commodities in response to a changing demand structure, and to enlarge the scale of the production unit. An important direction of agricultural development policy suggested by the Law was to foster the growth of family farms to 'viable units' that could earn income from agricultural production comparable to the level of non-farm household income. In order to improve farming efficiency, it was considered essential to increase the scale of the farm operation by promoting both a reduction in the numbers of inefficient farm units and cooperative operations among the remaining farms.

Despite the effort for structural adjustment, the rate of agricultural productivity growth was not raised sufficiently to prevent the rural–urban income gap from widening further. The reaction of farmers was to organize political lobbying for government support on agricultural product prices. Their pressure focused on rice, and resulted in 1960 in a rice price determination formula called the 'Production Cost and Income Compensation Formula'.

In this formula, the price of rice was determined by the cost of production at the paddy field in which yield per hectare was lower than the national average by 1 σ (one sigma, or one standard deviation). Since rice yield per hectare is in general inversely correlated with the cost of production per unit of output, this formula implies that the price thus determined covers the cost of production of marginal producers ('marginal' in the sense of a yield lower than average by about 1 standard deviation). A critical point in this formula is that wages for family labour are valued by non-farm wages in order to guarantee 'fair returns' for the labour of rice producers.

With this formula, the producer price of rice rose rapidly, corresponding to the rise in industrial wages, until 1968 when the accumulating surplus rice in government storage put a temporary brake on further price increases. From 1960 to 1968, the producer price doubled and the difference between the producer and the import prices widened from less than 50 to more than 120 per cent.

The Production Cost and Income Compensation Formula was designed to reduce the gap between farm and non-farm income and wages by raising rice prices. This policy appears to have achieved its goal. Income per agricultural worker compared to the income of manufacturing workers improved after 1960 with the rapid rise in agricultural relative to manufacturing prices. Increase in the price of rice – which constituted about 40 per cent of the total value of agricultural output before 1970, see Figure 2.2, p. 31 – was a major factor in improving the domestic terms of trade for agriculture. The rise in agricultural prices, together with increase in off-farm income, resulted in a marked reduction in the gap in income per person and per household between agriculture and non-agriculture (Column (9), Table 2.1, p. 20).

Satisfaction of the income parity objective involved substantial loss of economic efficiency. High rice prices should have reduced consumer surplus, not only by contracting the demand for rice itself but also by obstructing the shift of resources from rice to other high-demand agricultural products such as livestock and vegetables. The support of rice prices also depressed the outmigration of agricultural labour to non-agriculture. More conspicuous inefficiencies were the rapidly accumulating surplus rice in government storage, and the multiplying deficit from the food control programme.

How could such a waste of resources and loss of economic efficiency be tolerated during the 1960s? Certainly, it was the powerful political pressure of *NOKYO* that achieved the handsome rise in agricultural price support. Although the farm population declined in the course of rapid industrial growth, electoral districts have changed little, leaving the political weight of farm votes intact. Rural districts have continued as solid conservative blocks, which the present government ruled by the conservative Liberal Democratic Party (LDP) could not dare to lose (Hemmi, 1982).

Why, then, were the powerful rural votes unsuccessful in raising the price of rice during 1950s? Why could not the equally strong (or stronger) political power of landlords before 1939 achieve a comparable support for agriculture?

This question can be answered only in terms of decline in the role of rice as a wage good for industrial development. The big spurt of industrial development since 1955 ushered the Japanese economy into a new era. Within 10 years real *per capita* income more than doubled, and approached the level of Western Europe. Both the industrial structure and the export structure came to be dominated by capital-intensive industries. Labour shortage became a secular feature of the economy after 1960, and the wage differentials were greatly reduced among different sizes of enterprises and between blue collar and white collar workers.

With the dramatic increase in income and wages of industrial workers – particularly low-income manual workers – their diet rapidly changed. Indeed, according to a survey by the Labour Ministry

(*Maigetsu Kinro Tokei Chosa*), average compensation per industrial employee deflated by the consumer price index doubled in the 1955–70 period (and it doubled again in the following decade and half). Before 1960, decline in the starchy food staples in the total caloric intake of the Japanese was mainly attributed to the decline of inferior grains such as barley. From 1960, the share of rice also began to decrease. The absolute *per capita* consumption of rice declined from the peak of 118 kilograms per year in 1962 to 95 kilograms in 1970 (and further to 80 kilograms in 1980). Rice as an element in the consumption expenditures of urban worker households declined rapidly, from 10 per cent in 1960 to 4 per cent in 1970, and to 2 per cent in 1980.

Meanwhile, because of the rise in capital intensity and the transformation of the industrial structure, increases in the cost of living and in wages became less critical for the international competitive power of Japanese industry. Rice was no longer a critical wage good for industrial development. Through the process of postwar economic growth Japan's economy reached a stage in which there is little danger of urban disruption, such as the Rice Riot in 1918, occurring in response to food price increases. In such a situation, industrial and commercial concerns found it advantageous to keep farmers as their allies, even at the expense of high food prices, against the tide of socialist and communist movements. This appears to be the reason why the political pressure of *NOKYO* was able to achieve the high rate of agricultural protection.

Clearly, during the 1960s the Japanese economy entered the stage in which agricultural policies were entirely dominated by consideration of the 'agricultural adjustment problem'.

3 Structure of Agricultural Protection

Throughout the period of postwar economic growth, agricultural policies in Japan have become dominated by consideration of how to adjust agriculture to rapid economic development. The problem of reallocating labour from the farm to the non-farm sector at a sufficiently rapid pace to maintain intersectoral income parity has been especially severe in Japan, because comparative advantage has shifted away from agriculture relative to other industrial countries owing to unbalanced and rapid growth in industrial productivity. If the adjustment were left to the market mechanism, its cost would have had to be shouldered mainly by agricultural producers. To combat the situation, farmers have organized political lobbying for agricultural protection as a means to pass on a part of the adjustment cost to the general public. Their demand, when coupled with declining countervailing power from the side of the non-farm population, has resulted in a level of agricultural protection among the highest in the world, as observed in Chapter 1 (Table 1.2, p. 6).

In this chapter, we try to outline the various policy instruments being used to achieve the high level of agricultural protection, and assess the goals (and the consequences) of those policies.

3.1 INSTRUMENTS OF AGRICULTURAL PROTECTION

Agricultural protection policies are defined as 'the policies to increase agricultural output and income through government intervention into agricultural product and input markets'. The agricultural inputs subject to the protection policies are limited to the 'private goods' tradeable in the market place, and exclude the 'public goods' utilized jointly by a large number of agricultural producers, such as major canals for irrigation and drainage. The supply of public goods for which a market does not exist (or fails to allocate resources efficiently) is the prime responsibility of government in any situation. The protection policies defined here are therefore limited to government interventions into the markets of private goods. Major instruments of agricultural protection include: (a) border protection, (b) direct supports on farm product prices, and (c) subsidies on agricultural production inputs.

3.1.1 Border Protection

In the past, a major source of criticism on the trade practice of Japan from its trade partners has been the quantitative restrictions on the imports of agricultural commodities. Indeed, the import quotas (IQ) on beef and oranges have been publicized as a symbol of the closed nature of Japanese agriculture. At present, 22 agricultural and marine products are subject to IQ, as enumerated in Table 3.1. This number is larger than 19 in France who maintains the largest number of IQ restrictions among EC member counties. Foreign criticism on the IQ protection is especially great because it violates the GATT Code (Article XI). Under pressure from trade partners, the number of agricultural commodities in the IQ list was reduced from 102 in 1962 to 58 in 1970, and to 22 in 1974, but since then there has been no reduction. Meanwhile, the number of non-agricultural IQ commodities went down from 32 in 1970 to 7 in 1974, and down to only 1 (coal) in 1986 when leather and leather footwear were shifted from the IQ to the tariff-quota protection list.

The reduction in IQ restrictions has been resisted not only from farm producers but from the holders of import licences, as well as politicians and bureaucrats who have leverage in the allocation of quotas. Large windfall profits associated with IQ licences can be inferred from sharp declines in marketing margin in bananas, lemons and grapefruit when they were removed from the IQ list, as shown in Column (3) of Table 3.2.

Besides the IQ restrictions, the imports of several agricultural commodities are controlled by the trade monopolies of governmental or semigovernmental agencies: rice, wheat and barley by the Food Agency, beef, butter and powdered milk by the Livestock Industry Promotion Corporation, silk by the Silk and Sugar Price Stabilization Corporation and leaf tobacco by the Japan Tobacco Corporation. Among industrial market economies, Japan relies most heavily on state-trading as a means of agricultural protection; this is partly

TABLE 3.1 Commodities subject to import quotas in Japan (as of May 1987)

Category and no. of products	*Commodities*
Under the jurisdiction of the Ministry of Agriculture, Forestry and Fisheries (22)[a]:	
Dairy products (3)	Milk and cream (fresh)
	Evaporated milk, etc.
	Processed cheese, etc.
Meats and processed meat products (2)	Beef
	Mixed meats and other processed products
Processed grain products (2)	Flour
	Flour meal and cracked barley
Fresh and processed fruit and vegetables (6)	Oranges, etc. (fresh)
	Oranges, etc. (temporary storage)
	Fruit juices and tomato juice
	Fruit purees and fruit pastes
	Processed pineapple products, etc.
	Tomato ketchup and tomato paste
Starches and sugars (2)	Dextrose, lactose, etc.
	Starch
Local agricultural products and seaweed (3)	Beans
	Peanuts (except for crushing for oil)
	Konnyaku roots and edible seaweed
Fish and shellfish (3)	Herring, cod, yellowtail, and some other fish as well as cod eggs (fresh, refrigerated, or frozen)
	Same as above (salted or dried)
	Scallops, scallop eyes, and cuttlefish (excepting *Monko* cuttlefish)
Others (1)	Other processed foods
Under the jurisdiction of the Ministry of International Trade and Industry (1):	
Coal (1)	Coal briquettes

Note:
a The number of commodities is shown in parentheses.

TABLE 3.2 Changes in the marketing margin corresponding to the abolition of the import quota (IQ) system for selected agricultural commodities in Japan

Commodity		*Banana*	*Lemon*	*Grapefruits*
Time when IQ was abolished		*April 1963 (1)*	*May 1964 (2)*	*June 1971 (3)*
Wholesale price (yen/kg):				
Before abolished	(1)	167	405	326
After abolished	(2)	119	157	177
Import cif price (yen/kg):				
Before abolished	(3)	55	142	131
After abolished	(4)	52	125	34
Marketing margin (yen/kg):				
Before abolished	(5)=(1)–(3)	112	263	145
After abolished	(6)=(2)–(3)	67	32	34
Rate of decline in wholesale price (%)	(((1)–(2))/(1))	29	62	46
Rate of decline in marketing margin (%)	(((5)–(6))/(5))	40	88	83

Source: Y. Kano, *Nihon yo Nogyo Kokka Tare* (Let Japan be an Agricultural State) (Toyo Keizai Shimposha, 1984) p. 130.

because state-trading does not violate the GATT Code and hence has not been subjected to serious multilateral negotiations (Johnson, Hemmi and Lardinois, 1985).

While quantitative restrictions are strong in Japan, border protection by means of tariffs and levies seems to be relatively modest compared with other industrial countries, especially EC members. According to a calculation made by the Japan Ministry of Agriculture Forestry and Fisheries (JMAFF, 1984) based on the tariff rates agreed upon in the GATT Tokyo Round Negotiation, the average tariff (and levy) rate on the value of imported agricultural commodities in Japan would amount to 8.6 per cent which is higher than 2.9 per cent in the US but lower than 12.3 per cent in the EC. The high average tariff rate in the EC reflects the fact that the major means of agricultural protection in the EC is the variable levy system.

In Japan, the variable levy of the EC type is not so commonly used. A somewhat similar system applies to pork, for which the basic tariff is 5 per cent *ad valorem*, but the differential tariff is applied when it becomes higher than the basic tariff; the differential tariff is determined as the difference between the target price (the middle point of the 'price stabilization zone') and the import cif price. Also (for sugar and silk) a certain proportion of the difference between the target and the import cif prices is levied by the Silk and Sugar Price Stabilization Corporation; when the difference become larger, the rate of levy is raised so that the domestic market price would not decline below the floor of the price stabilization zone.

Several agricultural and agriculture-based commodities (including natural cheese, oats, corn for starch and industrial uses, malt, feeder cattle, fish meal, and leather) are protected by the tariff quota system. In this system, the primary tariff rates are applied to imports up to certain quantities and the elevated rates are levied from the imports exceeding those quantity quotas. In the case of natural cheese, dairy processors are allowed to import 2 kilograms of foreign cheese free of duty for every 1 kilogram of domestic cheese they purchase for producing processed cheese. Imports above this limit are subject to an *ad valorem* duty of 35 per cent.

Japanese markets are known to be open for the imports of feed grains. However, the duty-free entry of corn and sorghum for feed is allowed only if they are purchased by bonded feed mills. Otherwise imports are subject to a specific duty of 15 yen per kilogram in the case of corn and an *ad valorem* 5 per cent duty in the case of sorghum. Imports of mixed feed are limited by a 15 per cent duty. Such regulations discourage on-farm feed mixing and protect domestic feed mills, many of which are owned by agricultural cooperatives. Here is a case in which agricultural cooperatives gain at the expense of livestock farmers.

In addition, there are several non-tariff barriers. For example, besides the IQ restriction on oranges and orange juice a government programme to promote the sale of domestically produced tangerine juice requires that all imported orange juice be blended with at least 40 per cent of the tangerine juice. Unlike beef, imports of live cattle are not subject to the IQ restriction. However, the capacity of animal quarantine *de facto* limits the imports of live animals.

Unlike the EC and the US, overseas dumping of surplus agricultural commodities by means of either export subsidies or food aid has been practised in Japan only exceptionally. During the 1969–74 and the 1979–84 periods, when surplus rice was accumulated in the government inventory, about 4 million tons in total were exported to

Korea, Indonesia and other developing countries on concessional terms such as deferred payments with low interest charges. Much less reliance on overseas dumping than other industrial countries reflects the fact that – even though the level of agricultural protection by other means has been higher in Japan – it has not been high enough to create much export surpluses under the conditions of rapid decline in comparative advantage in agriculture.

3.1.2 Price Support

In addition to indirect supports by border protection, various agricultural products are under the direct price support of government. The largest and the most complicated price support programme applies to rice, and will be explained separately below. Wheat and barley produced domestically are purchased by the Food Agency within JMAFF, if their market prices decline below floor prices.

So-called 'Price Stabilization Programmes' for meat, dairy products and silk involve buffer stock operations to support domestic wholesale prices between certain ceiling and floor prices. These 'Price Stabilization Zones' are generally set higher than market equilibrium prices so, that deficits are generated from the programmes. These deficits are financed partly by levies on imports, and partly by transfer from the general budget. The same applies to the government purchase at floor prices of sugar cane, sugar beet and potatoes for starch-making. These price support and stabilization programmes are operated by the Livestock Industry Promotion Corporation and the Silk and Sugar Price Stabilization Corporation.

Deficiency payments from the government apply to a limited number of products such as soybean, rapeseed and milk for processing. In the case of soybean and rapeseed, the government payments to producers for the differences between the target and the market prices are financed totally from the treasury. In the case of milk for processing, the payments are partly financed by the profit of the Livestock Industry Promotion Corporation from state-trading in certain dairy products, and a part is financed indirectly by the revenue from state-trading (to the extent that the revenue is used to support the domestic prices of dairy products so that dairy manufactures can purchase milk from farmers at higher prices). Therefore, so long as the state-trading revenue was large, the deficiency payment scheme for milk for processing could be operated at a modest cost to the treasury. In more recent years, however, as the domestic production has been stimulated to expand by high prices and the imports declined, the treasury cost needed to maintain the high target price of milk has been cumulatively increasing.

A variation of the deficiency payment scheme used in Japan is the 'Price Stabilization Fund', to which government and producers pay contributions and from which producers receive deficiency payments if market prices decline below target prices. This scheme is applied to calves, vegetables and fruits for processing.

3.1.3 Subsidies

Japanese agricultural policy depends heavily on subsidies. As shown in Table 3.3, agricultural subsidies amounted to 49 per cent of total agricultural budget in 1960, and it increased further to 62 per cent in 1984. If we add to it a transfer from the General Account to the Food Control Special Account (which is essentially a subsidy on rice prices), the ratio of subsidy to total agricultural budget is as high as 80 per cent.

It is well known that Japanese agricultural subsidies are characterized

TABLE 3.3 Structure of the national government budget for agriculture in Japan

Budget item	*1960*	*1970*	*1980*	*84*
Total agricultural budget[a] (billion yen)	139	885	3108	2810
Share of agricultural budget to total budget (%)	7.9	10.8	7.1	5.5
Composition of agricultural budget[b] (%)				
Personnel, facility, logistics, etc.	21	9	9	10
Subsidy payments	49	46	61	62
Transfer to special accounts	30	45	30	28
(Transfer to food control special account)	(11)	(36)	(21)	(18)

Notes:
a Including supplementary budgets.
b Based on the composition of the initial budget of the Ministry of Agriculture, Forestry and Fisheries.

Source: Japan Ministry of Finance, *Zaisei Tokei* (Financial Statistics). JMAFF, *Nogyo Hakusho Fuzoku Tokei* (Statistical Appendix to the Agricultural White Paper).

by a large number of items and a relatively small disbursement per item. According to a detailed study by Imamura (1978), in 1977 the number of subsidy items under the jurisdiction of JMAFF was 592 (the largest among the ministries of the central government), but the average disbursement per item was only 2.4 billion yen or less than half of all the ministerial average of 5.1 billion yen. Such a large number of small subsidies has been considered essential for mobilizing a large number of small farmers for various policy programmes. Heavy reliance on subsidies has been a secular feature of Japanese agricultural policy since the interwar period when the 'poverty problem' was predominant (see Chapter 2).

It is difficult to ascertain how much the government subsidies contributed to the reduction in the cost of agricultural production to domestic producers and hence contributed to agricultural protection. However, as shown in Table 3.4, in 1984, 40 per cent of capital formation in agriculture was financed from government subsidy and 20 per cent financed from institutional loan. The institutional loan consisted of low-interest loans from the government fund channelled through the Agriculture, Forestry and Fisheries Credit Corporation, and loans from the credit departments of the agricultural cooperatives for which a part of interest is subsidized by the government. If the credit subsidy is added to the direct subsidy for agricultural investment, the total would amount to about one-half of agricultural

TABLE 3.4 Sources of finance for agricultural investments in Japan

Source	*1960*	*1970*	*1980*	*1984*
Total agricultural investment (billion yen)	457	1377	4034	4103
Composition (%):				
Government subsidy	15	24	39	40
Institutional loan	11	22	22	20
Private fund	74	54	39	40
Investment in land infrastructure (billion yen)	100	407	1766	1743
Composition (%):				
Government subsidy	66	71	80	82
Institutional loan	27	20	18	16
Private fund	7	9	2	2

Source: JMAFF, *Nogyo Oyobi Noka no Shakai Kanjo* (Social Accounts of Agriculture and Farm Households).

capital formation in Japan, since the interest rates of institutional credits are one-third to two-thirds lower than the market rates.

The ratio of subsidy to capital formation is especially large in the case of investment in land infrastructure. It may appear that (considering the public-good characteristics of land infrastructure such as irrigation and drainage facilities) subsidies allocated to it should not be counted in the cost of protection. It must be recognized, however, that the portion of land infrastructure investment allocated to major canals and other overhead facilities is only about 20 per cent and the rest is allocated to farm ditches and farmland consolidation and reshaping, for which individual beneficiaries can be easily specified.

It should be noted in Table 3.4 that the weight of government subsidies as a source of finance of agricultural investment continued to increase rapidly from 1960 to 1980. This paralleled the rise in the nominal rate of agricultural protection in Table 1.2 (p. 6), indicating that agricultural protection was strengthened simultaneously by means of both subsidies to input costs and supports on product prices in the process of postwar economic development.

How does the level of agricultural subsidy in Japan compare with those of other industrial countries? Table 3.5 presents an attempt at a comparison between Japan and EC member countries. First, in terms of the ratio of current subsidies to agricultural GDP, Japan's level in 1983 (9.0 per cent) was more than 60 per cent higher than the EC average (5.4 per cent). Moreover, while this ratio remained stable or slightly declined in the EC, it rose sharply in Japan during the 1975–80 period.

Second, in terms of the ratio of capital subsidies to agricultural fixed capital formation in 1980, Japan's level (41.1 per cent) was about 50 per cent higher than the EC average (30.4 per cent). The higher rates of both current and capital subsidies reflect the more severe agricultural adjustment problem corresponding to comparative disadvantage in agriculture in Japan relative to other industrial countries. At the same time, the high rate of capital subsidies in Japan might be explained partly by the specific structure of Japanese agriculture, in which a large number of very small-scale farms grow rice mainly in the fields irrigated by gravity systems. Such structure requires larger public relative to private investment in Japanese agriculture than in western agriculture, because even relatively small irrigation and drainage facilities are used jointly by many farmers, thereby having the characteristics of public goods.

In terms of the ratio of total subsidies to agricultural GDP in 1980,

TABLE 3.5 Comparison of agricultural subsidies from central governments to Japan and EC member countries[a]

	Ratio of					
	Current subsidies to agricultural GDP				*Capital subsidies[b] to agricultural fixed capital formation[c]*	*Total subsidies[d] to agricultural GDP*
	1975	*77*	*1980*	*83*	*80*	*80*
Japan	3.1	3.8	9.1	9.0	41.1	32.9
France	5.9	5.4	3.8	3.7	30.5	10.4
Germany (FR)	5.9	9.2	7.4	4.4	15.7	13.3
UK	14.8	5.5	6.9	10.4	30.4	14.9
EC average	5.9	5.8	5.1	5.4	30.4[e]	12.5[e]

Notes:
a Includes direct subsidy payments from the central government to farmers but does not include subsidy payments to local governments for their activities to facilitate agricultural production.
b Capital subsidies for EC countries are national expenditures for modernization of farms, land improvement, improvement of production potential, development of less-favoured areas, national disasters, and development of rural areas.
c Agricultural fixed capital formation for EC countries includes fixed capital formation in forestry and fisheries.
d Current subsidies plus capital subsidies.
e Includes aids from the European Agricultural Guidance as Guarantee Fund Guidance Section.

Source: EC Commission, *The Agricultural Situation in the Community*, EUROSTAT *Basic Statistics of the Community*; OECD, *National Accounts*; JMAFF, *Nogyo oyobi Noka no Shakai Kankyo* (Social Accounts of Agriculture and Farm Households).

Japan's level (32.9 per cent) was almost double the EC average (12.5 per cent). In Japan, this ratio rose progressively from 14.0 per cent in 1970 to 14.9 per cent in 1975, to 33.6 per cent in 1980, and then fell slightly to 32.8 per cent in 1984. Judging from this trend, together with the international comparison in the ratios of current subsidies, the level of agricultural protection in Japan by means of government subsidies is considered to have risen relative to other industrial countries, together with border production and price support.

While Japanese agriculture is heavily subsidized, the tax burdens shouldered by farm producers seem much lower than that of the non-farm population. According to a study of Ishi (1981), for the

1970–8 period the rate of capturing taxable income – the ratio of income captured as taxable by the tax office to real taxable income – was only 20–30 per cent for agricultural producers, whereas the ratios were 60–70 per cent for the non-farm self-employed and 90–100 per cent for employees. A study by Honma (1987) shows that the tax rate on agricultural income in 1983 was one-half the rate on non-agricultural income within farm households. Further, he estimated that the ratio of subsidy receipt by farmers to their tax burden was less than 30 per cent in 1983, and that the net subsidy (subsidy receipt minus tax payment) to agriculture amounted to 35 per cent of total agricultural GDP. Such relation between tax and subsidies on agriculture today is diametrically different from the relation which prevailed in the prewar period, when the agricultural sector was heavily taxed relative to the non-agricultural sector as well as to its own subsidy receipt (refer back to Tables 2.4 and 2.5, p. 40).

3.2 GOALS AND CONSEQUENCES OF THE RICE POLICY

Rice has traditionally been the most important product in Japanese agriculture, as well as the most important food item to the Japanese. How to control the supply of rice so as best to serve the purpose of solving the agricultural problem in each stage of economic development has always been the major policy agenda. Today, rice is subject to the strongest protection among agricultural products, reflecting the severe agricultural adjustment problem. In order to understand the mechanism of agricultural protection in Japan we should examine how various policy instruments have been combined for the support of the domestic rice sector, and what consequences have been produced.

Readers who have the background of economics may wish to read this section with reference to an analytical model in Appendix B, p. 132.

3.2.1 Mechanism of Price Support

Since the Second World War rice has been under direct government control, based on the Food Control Law of 1942. Initially, this law was designed to ration distribution of all staple food items during the severe shortages of the war. As the food supply recovered, most food items were, one after another, removed from direct government control. Since 1952 (when wheat, barley and naked barley were shifted from direct to indirect control based on state-trading and government purchase at minimum guarantee prices to producers), rice has been the only commodity which remains under direct control in its distribution.

Initially, the whole marketing process of rice from producers to consumers was under the direct control of the Food Agency, and prices were regulated from the farm-gate to the retail level. Those regulations were relaxed gradually. In 1972, the control on channels from wholesalers to consumers was removed, and retail pricing became free. Since then, the direct control has applied only to a marketing chain from the farm-gate to the wholesale level.

In the course of economic development that followed Japan's postwar recovery, the direct government control of rice distribution changed its role from protecting consumers to protecting producers by means of price support. The price of rice was raised especially fast during the 1960s after the introduction of the Production Cost and Income Compensation Formula in rice price determination, and diverged progressively from the international level (refer back to Table 1.3 and explanations in Chapter 2, p. 9). Needless to say, the high level of price support for domestic producers has been associated with the restriction on rice imports that are monopolized by the Food Agency. Since the mid-1960s rice imports have been virtually nil (except for a small amount of glutinous rice for sweets).

During the 1960s, the price of rice was raised not only far above the world price but also above the market equilibrium price under autarky. As the result, domestic production expanded in excess of consumption, resulting in an accumulation of surplus rice in government storage. The accumulated surplus pressed hard on the current expenditure of the Food Control Special Account in the form of increased storage cost (see Table 3.6).

Moreover, as the inventory was carried over years, the quality of rice declined and had to be disposed of for uses other than domestic food. The sale prices of old rice for exports and for feed and industrial uses in the surplus-rice-disposal programme were at best half (and at worst only one-seventh) of the government sale price for normal domestic consumption. Because exports of rice were made on concessional terms (for example, with no interest charge for the first 10 years and only 2–3 per cent interest thereafter), the real price of exports may well have been less than half the nominal price, which itself was less than half the domestic producer price. Moreover, Japan was forced to limit rice exports because of strong objections

TABLE 3.6 Changes in the cost of the food control system in Japan

	Total general budget	*Total agricultural budget*[a]	*Food control cost*			*Share of food control cost in*	
			Deficit of Food Control Special Account	*Cost of acreage control*	*Total*	*Total budget*	*Agricultural budget*
	(1)[b]	*(2)*[b]	*(3)*[b]	*(4)*[b]	*(5)=(3)+(4)*[b]	*(5)/(1)*[c]	*(5)/(2)*[c]
1960	1765	166	29	–	29	1.6	17.4
1965	3744	404	120	–	120	3.2	29.8
1970	8213	992	374	81	455	5.6	46.0
1975	20 837	2289	811	106	917	4.4	40.1
1980	43 681	3776	652	304	956	2.2	25.3
1985	53 223	3390	456	239	695	1.3	20.5
1986	53 325	3230	364	250	614	1.1	18.7

Notes:
a Includes budgets for forestry and fisheries.
b Billion yen.
c (%).
Source: JMAFF *Norin Suisan Yosan no Setsumei* (Explanations of the Budgets for Agriculture, Forestry and Fishery).

from rice-exporting countries. The price of rice for industrial use was relatively favourable, but the demand for this purpose was very limited. The demand for feed use is elastic only at a very low price on a par with feedgrain prices. The average salvage value of surplus rice thus tends to decline sharply with increasing disposals of surplus rice, because a larger proportion of the disposal must be allocated to feed use. Capital loss resulting from the surplus-rice-disposal programme, which is included in the deficit of the Food Control Special Account, thus becomes very large when old rice stock accumulates.

In order to prevent the accumulation of surplus rice, the control on rice acreage has been introduced since 1969. The acreage control also requires a substantial cost in the form of subsidy payments for withdrawing paddy fields from production or diverting their use to non-rice crops. As rice production technology has improved while consumption has declined steadily since the early 1960s when rice became an inferior good, the acreage control has had to be strengthened. In the mid-1980s, the production capacity of rice with no acreage control is estimated to be about 14 million tons, while the consumption is a little over 10 million tons. Since the average yield in normal years is about 5 tons per hectare, 70 to 80 million hectares (or nearly one-quarter of total paddy field area in Japan) must be set aside from production.

In addition to the costs of surplus disposal and acreage control, the deficit of the rice control programme in Japan inevitably increases, according to the increase in rice output because the margin of the government's rice marketing is negative. During the 1960s, when the government yielded to farmers' political pressure for raising the price of rice at which the government purchased from producers (the so-called 'producer price'), the government sale price to wholesalers (the so-called 'consumer price') was not raised so fast in consideration of consumers' resistance. At that time, tax revenue increased very fast corresponding to the rapid growth in national income and, hence, it was cheaper politically to support farmers by tax revenue than by higher prices to consumers. Since then, the negative marketing margin to the Food Agency in its distribution of rice has remained a secular feature of the food control system (*Shokkan Seido*), despite mounting pressure to correct it from the need to reduce the national budget deficit that has become serious since the first Oil Crisis of 1973. The structure of the negative margin is illustrated in Figure 3.1 for the case of 1986. The government purchase price (311 000 yen per ton) was lower than the sale price (310 000 yen) by 1000 yen per ton. Adding to it the government administration cost (61 000 yen) created to the negative margin of 62 000 yen per ton of rice distributed through the Food Agency.

Given the negative margin, increases in rice output and marketable surplus automatically result in an increase in the deficit of the Food Control Special Account, so long as all rice must be distributed through the government channel. A device designed to reduce this deficit was the 'voluntary-rice' (*Jishu Ryutsu Mai*) channel that was introduced in 1969 into the food control system in addition to the pre-existing 'government-rice' (*Seifu Mai*) channel. In the latter, the Food Agency purchases rice from producers through agricultural cooperatives and sells it to wholesalers. In the former, cooperatives sell directly to wholesalers at a price negotiated between representatives of cooperatives and wholesalers. Rice marketed through the voluntary-rice channel is limited to high-quality rice that can command a price higher than the government-purchase price. The total quantity of rice that each producer can sell through either (or both) the voluntary or the government channel is limited to a certain quantity according to acreage quota determined by the acreage control programme.

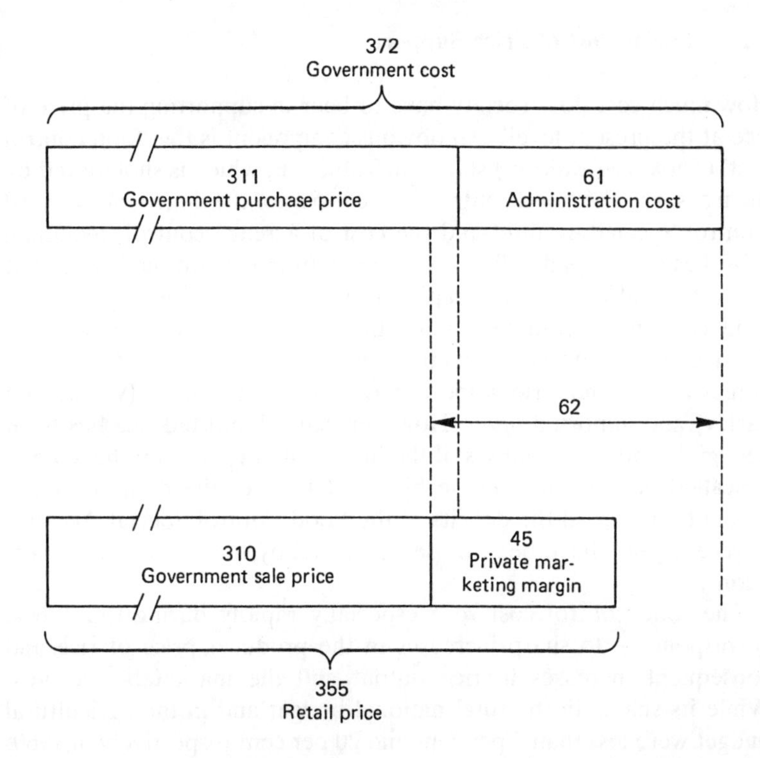

FIGURE 3.1 The structure of rice prices (000 yen per ton of brown rice) in Japan (average prices of the first and second grades including packing costs, 1986)

In order to increase the share of voluntary rice, the government gives subsidies to producers for their rice sold through the voluntary-rice channel. The share of voluntary rice has grown in 1987 to occupy about one-half of the 7 million tons of rice distributed through the two official channels in the food control system. Besides rice passing through those official channels, it is said that about 1 million tons are marketed through illegal private channels, commonly called 'free-market rice' (*Jiyu Mai*). The government has tried hard to prohibit this illegal rice distribution because its increase invalidates acreage quotas, with a resulting increase in surplus production. In practice, however, it has been increasingly difficult for the government to enforce regulations on rice distribution – which originated in the midst of hunger and destitution during the war – in a situation of affluence and varied preferences among the Japanese people.

3.2.2 Social Cost of Price Support

How much cost does society have to bear in supporting the price of rice at the present level? An obvious component is the 'food control cost' (*Shokuryo Kanrihi*) shown in Table 3.6, which is shouldered by the treasury. The 'food control cost' consists of the deficit of the Food Control Special Account and the cost of acreage control, including subsidies for the paddy fields withdrawn from cultivation. The deficit is the total of current and capital losses from the operations of the food control programmes, plus subsidies to 'voluntary rice'. The Food Control Special Account includes the accounts of not only domestic rice but also imported rice, domestic *mugi* (wheat and barley) and imported *mugi*. However, since imported rice has been negligible and the surpluses of the imported *mugi* account have been cancelled out on the average by the deficits of the domestic *mugi* account, almost all the deficits of the Food Control Special Account in recent years have been those generated by the rice control operations.

The food control cost rose especially rapidly during the 1960s, corresponding to sharp increases in the producer price of rice and consequent increases in rice output and the marketable surplus. While its shares in the total national budget and in the agricultural budget were less than 2 per cent and 20 per cent respectively in 1960, they exceeded 5 per cent and 40 per cent in 1970. Since then, those shares have been declining due to the efforts of curbing the deficit by such means as the promotion of acreage control and voluntary rice. Still, in 1985, it amounted to about 700 billion yen, close to 4 billion US dollars.

The food control cost paid from the treasury is only a part of the total cost the society has to bear for the support of domestic rice prices. A part must be shouldered by consumers, to the extent that they pay a price higher than the international price. The consumer cost of the rice price support can be roughly estimated as follows.

In the mid-1980s, the total amount of rice sold from producers per year – which was about equivalent to the total quantity of consumption – is estimated to have been about 8 million tons. While the government sale price to wholesalers was about 300 000 yen per ton, it is estimated that the import cif price was about 120 000 yen. If it can be assumed that any amount of rice can be imported at this price (the 'small-country' assumption), it is considered that the burden of consumers would have amounted to as much as 1400 billion yen, which is obtained by multiplying the difference between the government sale and the import cif prices (300 000 – 120 000 = 180 000 yen) with the quantity of market supply (8 million tons).

In reality, the small-country assumption would not hold, and the international price should have risen if the rice imports were liberalized. Even in the case of complete trade liberalization, however, it is not possible for the import price to exceed the domestic equilibrium price in Japan under autarky. The autarky equilibrium price is estimated as about 170 000 yen (Otsuka and Hayami, 1985). Therefore, the lower bound of the consumer cost is obtained by multiplying the difference between the government sale and the autarky equilibrium price (300 000 – 170 000 = 130 000 yen) with 8 million tons; this amounts to about 1000 billion yen.

Since the budget cost for the rice price support was about 700 billion yen in 1985, the total social cost involved in the price support programme in the mid-1980s would have been within a range from 1700–2100 billion yen, which amounted to as much as 40–70 per cent of total farm income from rice production (about 3000 billion yen). In addition, much of the production subsidies were paid to the rice sector. Therefore, it is likely that well over one-half of rice-farming income was really a transfer from consumer-cum-tax payers to producers.

Not the whole social cost of the price support programme is transferred to producers; a part is wasted in the production and the marketing processes. It is estimated that in 1980 the net loss in social economic welfare as measured by 'deadweight loss' amounted to 781 billion yen, or nearly 40 per cent of the total social cost (Otsuka and Hayami, 1985).

3.2.3 State and Market

The rice price support programme involves a highly complicated system consisting of various policy instruments which are often mutually contradictory, such as stimulus to production by means of high prices and restraints on production by subsidizing for the withdrawal of paddy fields. The complication has multiplied over time as state intervention in market forces has created unexpected problems through the response of market forces to the intervention policies, necessitating additional interventions.

The dynamism of policy changes resulting from interactions between market forces and state intervention is illustrated in Figure 3.2.

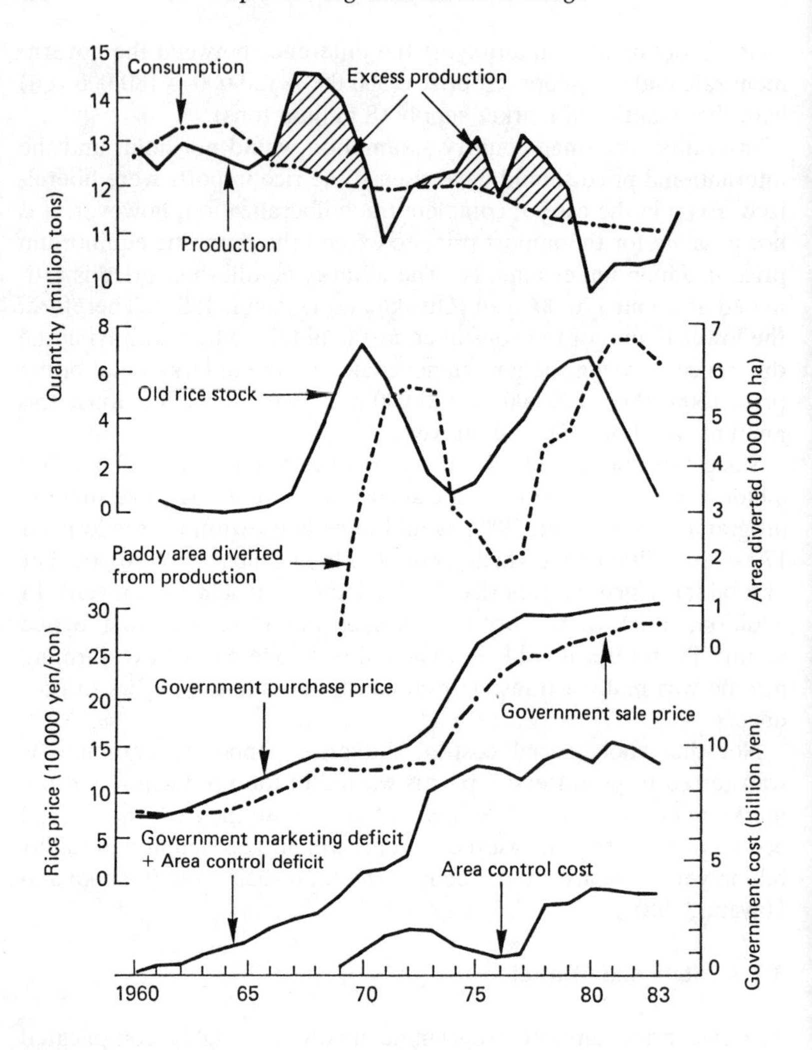

FIGURE 3.2 Changes in the rice market and in the government financial burden due to rice support programmes

Source: Japan Ministry of Agriculture, Forestry and Fisheries (JMAFF), *Shokuryo Kanri Tokei Nenpo* (Annual Report of Food Control Statistics); *Norin Suisan Sho Tokei Hyo* (Statistical Yearbook of JMAFF); *Shokuryo Jikyu Hyo* (Food Balance Sheets); *Nogyo Hakusho Fuzoku Tokei Hyo* (Statistical Appendix to the Agricultural White Paper)

First, after the introduction of the Production Cost and Income Compensation Formula in 1960, the government purchase price was raised very rapidly. The deficit was created because the government-sale price was raised with a few years' time lag.

However, the extremely rapid increase in the government deficit during the 1960s, as shown in Table 3.6 (Column 3), resulted more from the response of the market to the increased price than from the increase in the negative government-marketing margin. The increased producer price stimulated production and market supply. From 1960 to 1968, total rice output increased by 4 per cent from 12.5 to 14.2 million tons, while total sale to the Food Agency rose faster by as much as 67 per cent from 6.0 to 10.0 million tons. Given the negative marketing margin, the deficit from the food control programme increased proportionally with the increase in rice procurement by the Food Agency.

More serious was the accumulating surplus in government storage. During the 1960s, rice became an inferior good corresponding to *per capita* income rises; average *per capita* rice consumption per year declined steadily from a peak of 118 kilograms in 1962 to 100 kilograms in 1970. In addition to the negative income elasticity of demand, the increased price should have also contributed to the decline in rice consumption to some extent. With the bumper crop in 1967, the excess supply of rice became especially evident in the form of a sharp increase in the quantities of old rice in the government stocks.

The multiplying financial burden arising from excess supply forced the government to introduce simultaneously three measures in 1968: (a) restraint on the price of rice, (b) acreage control, and (c) disposal of surplus rice. For the subsequent three years, the producer price was frozen. During those years, the wage rate in the non-farm sector continued to rise. Therefore, if the Production Cost and Income Compensation Formula were strictly followed, the government-purchase price should have been raised. The fact that the price was fixed despite the Formula which guaranteed the return to farm labour at a rate equivalent to that of non-farm labour shows that this principle was applicable only within a limit of financial tolerance. The financial ceiling was elevated greatly during the 1960s, due to a large increase in tax revenue corresponding to high economic growth. It appears that the rapidly increasing financial burden from the price support, augmented by the response of the market, caught up the ceiling by the end of the 1960s.

In the short run, both demand for and supply of rice are inelastic and, therefore, the excess supply does not become significant even if the price is raised above a market equilibrium. In that situation – irrespective of how much the producer price would be raised – it would not cause much financial burden to government if the consumer price were raised in parallel even with a few years' time lag.

In the long run, however, demand and supply do not remain price inelastic. The increased price stimulates not only application of fertilizer and other current inputs, but also long-term investments such as irrigation, agricultural research and development. When the excess supply is created as the result of long-run adjustments in demand and supply to the price support, the resulting costs such as storage and surplus disposal costs can no longer be passed on to consumers.

During the 1960s, average *per capita* income rose very rapidly, at the rate of doubling in a decade in real terms, and rice consumption declined both absolutely and relative to total household consumption expenditure. The increase in the consumer price of rice was therefore not so strongly resisted. In the short run, the rice price support could be raised without too much stress on the treasury, since it was possible to pass on a large part to consumers. But when it met the long-run response of the market, it created an unbearable burden to the government.

It may appear strange to see that the rice price began to be raised again in 1973, as soon as the disposal of surplus rice was completed and the demand–supply equilibrium was restored by the success of the acreage control programme. This was partly due to the outbreak of the so-called 'World Food Crisis' of 1973–5. Sharp increases in world food prices, coupled with the US soybean embargo, stirred up anxiety in the public mind. The farm bloc took advantage of this situation in their lobbying for the price increase by advocating greater food self-sufficiency and security.

However, a more basic factor underlying the second surge of the rice price support seems to be that policy makers failed to see latent market forces ready to create a large surplus again in response to further price increases. Even if they foresaw this, it might have been difficult for them to present sufficiently strong evidence to counter the pressure for the price increase. Or it might have been the case that, given the politicians' high rate of discount for future costs and benefits arising from their immediate need for staying in office, it was to their advantage to yield to the pressure from the farm bloc. The truth would probably be a combination of these hypotheses.

At any rate, the price support was raised and the acreage control programme was relaxed. There followed a repetition of the experience of the 1960s. With increasing excess supply, surplus stock was accumulated and the government deficit was escalated, which made a price freeze and a strengthening of acreage control inevitable in the late 1970s.

The policy-induced cycles thus created have involved a large waste of resources. An obvious example is the accumulation of surplus rice and its disposal at a huge cost. Another example is that the paddy fields (which had been converted from upland fields with large investments for the installation of irrigation systems in response to high prices) were forced to be diverted from rice production to upland crops just when the construction was completed. Those examples illustrate how much social loss became inevitable from state intervention, when they were made without due consideration of the response of market forces.

3.3 THE STALEMATE OF PROTECTION POLICIES

By now, readers should be able to appreciate both the complexity and the intensity of agricultural protection policies in Japan. It is typical (but not at all exceptional) of the rice policy that state intervention in market mechanisms tends to magnify as the response of market forces creates unexpected problems; those new problems necessitate additional intervention, while the pre-existing intervention policies are hard to remove because they are locked in the vested interests that they themselves have produced.

It has been pointed out that (in the case of IQ commodities such as beef) not only traders holding quota licences but also government officials, administrators and politicians who enjoy considerable power in allocating these licences strongly resist the abolition of the IQ system. In the case of the rice policy, the same level of producer protection could have been achieved at a lower cost if the present system of direct control had been shifted to an indirect control system (similar to that of wheat and barley) based on state-trading and minimum guarantee prices. However, such a move has been strongly resisted by agricultural cooperative associations (*NOKYO*) which, under the present system, are designated as a sole marketing agents from the farm-gate to the wholesale level and need no major effort to earn large commissions. Their camp has been joined by some traders who are designated by the Food Agency as official agents to handle legal rice at the wholesale level. It is also well known that the allocation of agricultural subsidies – especially to land-infrastructure projects of which major beneficiaries are the contractors who often dominate local politics – is a major source of power for both bureaucrats and politicians.

Those agents, both private and semi-public (*NOKYO* in particular), who enjoy institutional rents created by the protection policies are indispensable vote-collectors for politicians as well as creators of lucrative employment opportunities for bureaucrats after their retirement. Thus, the 'iron triangle' formed by *NOKYO*, JMAFF and the ruling LDP (Hemmi, 1982) stands ready to block any move to reduce the institutional rent, even in the cases where such reforms match the interests of farmers themselves.

As explained in the previous two chapters, rapid growth in agricultural protection since the mid-1960s has been based on the decline in domestic countervailing power against protectionism, as well as the increase in demand for protection policies from the farm bloc corresponding to the loss in comparative advantage in agriculture. Growing affluence (and the resulting decline in the share of food in total consumption expenditure) have raised the ceiling of consumers' tolerance to high food prices. Protection policies have therefore seldom met strong organized protests by consumers, trade unions and business concerns.

In such a situation, two forces have been operating to impose a limit on the growth of agricultural protection in Japan. One is, of course, fierce protests from abroad, mainly food-exporting countries. Another is the response of domestic market forces. It has been observed that the increase in the support of rice prices stimulated production to such an extent that a further increase in price support became difficult as it came up against the budget ceiling. Equally critical is the response in demand. For example, the imports of processed food products, such as spaghetti and *arare* (rice cookies), have been increasing so rapidly as to imperil the domestic processing industries because the trade liberalization of product imports has not been accompanied by a similar liberalization in raw material imports. If the present trend continues, many food-processing industries will be forced to move their production base abroad where cheap raw materials are available, with the result of a serious shrinkage in an already slumping demand for domestic agricultural products.

Thus, to use the terminology of Albert Hirschman (1970), in the absence of internal 'voice' (political protests), the 'exit' of customers (market response) is now operating as a brake on further growth in agricultural protection in Japan, together with the strong 'voice' from abroad. The constraint imposed by the internal exit and the external voice has been felt increasingly in recent years as food consumption has come closer to saturation while Japan's comparative advantage in manufacturing has become more visible with an accumulating trade surplus and/or appreciating yen. At the same time, the deceleration in the growth of tax revenue and a resulting national budget deficit corresponding to a shift of the Japanese economy from the high-growth to the low-growth regime since the first Oil Crisis of 1973 has further strengthened the constraint.

Yet the complexity of agricultural protection policies has continued to multiply through interactions among state intervention, market forces and vested interests reinforcing each other. Meanwhile, there has been little progress in structural adjustments needed to upgrade the productivity of Japanese agriculture. A large number of inefficient mini-sized farms incompatible with the introduction of modern labour-saving technologies has been preserved, especially in the rice and beef sectors to which the strongest protection has been given. Investment in the education of rural people and in agricultural research and development, (which should have been more effective in assisting farmers) has been too little relative to their returns, because a major portion of government budget has been eaten up by the protection policies. Increasingly, Japanese agriculture has been locked into a stalemate that is an inexorable consequence of its protection policies.

4 Failure of Structural Policies under the Agricultural Basic Law

If the increase in demand for agricultural protection in the course of postwar economic development stemmed from the shift in comparative advantage from agriculture to industry, was it not possible to reverse the trend by accelerating the growth rates of agricultural productivity through structural adjustments? In fact, ever since the 1961 Agricultural Basic Law was enacted with the goal of making agricultural income levels equal to those in other industries, the Japanese government has worked untiringly to achieve this target through improving the agricultural structure. As indicated by the 'selective expansion' slogan, this has meant policies designed to raise agricultural production efficiency and farm income by transferring resources from the production of farm products of low-income elasticities to those of high-income elasticities, and by expanding the scale of operations. Yet despite these policy efforts, it proved difficult to achieve income equalization through agricultural restructuring alone, and protective policies had to be resorted to, primarily because the rapid economic growth induced such rapid increases in non-agricultural income that agricultural restructuring and the improvements in labour productivity in agriculture could not keep pace. But was it not possible to achieve structural adjustments at a sufficiently rapid rate so that the income parity could have been achieved through an increase in agricultural productivity? What were the major impediments to the structural adjustments?

4.1 THE BOTTLENECK OF PRODUCTIVITY GROWTH

A chief restraint on raising labour productivity in Japanese agriculture has been the small size of farms. Japan's average farm size is still

TABLE 4.1 International comparisons in labour and land productivities, and in land–labour ratios in agriculture, 1980 (1978–82 averages)

	Agricultural output per		*Area per male worker*		*Area per farm*	
	Male worker[a]	*ha. of agricultural land*[a]	*Agricultural land*[b]	*Arable land*[b]	*Agricultural land*[b]	*Arable land*[b]
USA	285	1.2	221	107	193	94
UK	116	3.1	36	14	66	25
France	102	4.1	23	14	25	15
Germany (FR)	114	6.0	18	11	14	9
Italy	48	5.0	5	4	9	7
Japan	28	12.2	1.7	1.5	1.2	1.1

Notes:
a Wheat unit, equivalent to 1 ton of wheat into which other agricultural products are converted according to the ratios of their prices relative to the price of wheat.
b Hectares
Source: Y. Hayami and V. W. Ruttan, *Agricultural Development: An International Perspective*, revised edn (Johns Hopkins University Press, 1985) Appendix A.

only a little over 1 hectare, and a very small percentage of farms have more than 3 hectares (see Table 2.3, p. 27). This scale of operation is well below those of other industrial countries, as shown in Table 4.1. Real agricultural output per worker in Japan is only about one-tenth of that of the US and one-half to one-quarter of that of EC members. Since Japan is surpassing other industrial countries in agricultural output per hectare of farm land, it is evident that the major factor underlying the comparatively low level of labour productivity in Japanese agriculture is the meagre endowment of agricultural land relative to labour force.

Before the mid-1950s when the 'high economic growth' of Japan began, the low productivity of labour did not necessarily mean weakness of Japanese agriculture in international competition. In that stage of economic development, labour was cheap relative to land and capital. Hence the cost of production was low, even if a large amount of labour was used per unit of land area, so long as the use of capital was effectively saved. Labour-using and land-saving technology geared to raising land productivity with large applications of labour and fertilizer per unit of limited land area was an efficient technology in the land-scarce and labour-abundant economy. Until the high-growth period, the agrarian structure of Japan (which consisted of a large number of small-scale but relatively homogeneous family farms) had worked efficiently, as it had been effective in facilitating the development and diffusion of land-saving technology and thereby contributed to the solution of the 'food problem' (see Chapter 2).

However, the traditional agrarian structure became inefficient as the wage rates were raised. In a high-wage economy, in order to reduce the cost of production, it is critically important to save the use of labour. For saving labour it is necessary to promote farm mechanization. In general, the larger is the scale of machinery used, the more effectively can labour be saved. To utilize large-scale machinery efficiently, farm size must be large.

Farms in Japan distributed with an average operational holding of about only 1 hectare, and more than 95 per cent of them operating on less than 3 hectares, are incompatible with the efficient use of large-scale machinery that is the carrier of modern labour-saving technology. On such small farms the scope of substituting capital for labour is severely limited, even though wage rates have increased sharply relative to farm machinery prices. It is true that by 1980 the level of farm mechanization – as measured by tractor horsepower per hectare of cultivated land – had risen to 8 hp in Japan, compared with 4 hp in France and 7 hp in West Germany. But, because the average area of cultivated land per worker in Japan in 1980 was only about one-tenth that of France and one-quarter that of Germany, tractor horsepower per worker in Japan remained only about one-quarter that of France and Germany (Hayami and Ruttan, 1985).

Therefore, in order to foster a 'viable agriculture' that is capable of yielding income per full-time farm worker at a level equivalent to the average income of non-farm workers without relying on government protection, it is vital to expand the scale of farm operation so that scale economies inherent in modern labour-saving technology can be exploited effectively. Expansion in farm size is only a necessary (but not a sufficient) condition to achieve viable agriculture. The traditional effort to raise land productivity by means of developing biological technology should not be neglected, since land productivity is an important component of labour productivity. Crop mix must also be changed so that the share of commodities for which there is demand increases in response to income growth. Above all, farmers' human capital must be upgraded so that they can reallocate resources efficiently in response to changes in technology and market conditions. Yet, unless the traditional agrarian structure, consisting of a large number of small farms, can be transformed into a structure dominated by a small number of large farms, it will be difficult to achieve the goal of viable agriculture, given the high wage rates in Japan comparable with those of other industrial countries.

4.2 AN UNREALIZED DREAM OF VIABLE FARMS

Policy makers in Japan were well aware of the need for structural change in agriculture corresponding to a shift from a low-wage to a high-wage economy. The 1961 Agricultural Basic Law set as a prime policy goal an equilibrium in income and standard of living between farm and non-farm workers (Article I). The Basic Law is a national charter that defines the basic direction of government policies on agriculture. Modelled after similar charters in Western Europe such as *Das Landwirtschaft Recht* in Germany (1955), the Agricultural Act in the United Kingdom (1957) and the *Loi d'orientation agricole* in France (1960), it aimed to specify the basic direction of agricultural adjustment in economic development and the policies necessary to achieve the adjustment.

The Agricultural Basic Law of Japan specifies that the farm – non-farm income parity objective should be achieved mainly by structural adjustments including, above all, expansion in operational farm size (Article III, Section 3), together with a shift of production to the commodities of rising demand (Article III, Section 1) and improvements in land infrastructure and agricultural technology (Article III, Section 2).

4.2.1 A Target of the Basic Law

The agrarian structure that the Basic Law aimed to establish was the one dominated by 'viable farm units'. The viable units are defined as 'farms sufficiently large in operational scale so that family members engaged in farming are able to work in their full capacity and earn high enough income to enjoy the level of living equivalent to that of non-farm workers' (Agricultural Basic Law, Article XV). In official statistics, farms that earn income per farm household member equal to (or above that of) non-farm employees who are living in rural areas are classified as 'viable units'.

The initial target of the government was to create 1 million viable farms by 1970. In order to achieve this goal, it was considered necessary that about one-quarter of a total of 6 million farms should, be closed down within 10 years, their land passing to viable units. Expectation of such a rapid decline in the number of farms was not a mere dream. In the decade before the Basic Law was enacted there had indeed been a sharp decline in the number of farmers' sons who planned to become farmers.

From the Meiji era until the Second World War, the total numbers of farms and gainfully employed workers in agriculture had remained roughly constant at the levels of 550 million farms and 1400 million persons, whereas both the number of farms and farm workers were abnormally inflated for the early post-war years due to repatriation to rural areas from devastated cities and overseas territories. The numbers of farms and workers had been maintained constant before 1939, because every year about 400 000 farm children newly graduated from schools had joined the agricultural labour force; a half of these school-leavers in agriculture had been males of whom about 75 per cent had succeeded and operated farms for a subsequent 30 years on the average.

As the data in Table 4.2 show, in 1950 the number of new school-leavers in farm households who began to engage in farming was 440 000, of which 230 000 were males. It can reasonably be assumed that 75 per cent of these males (or 175 000 farm boys) intended to succeed their parents' job. Suppose that the heads of 618 million farm households existing at that time had to be replaced within 30 years. Then, for the number of farms to remain constant, 206 000 farm boys fresh from schools with an intention to succeed their fathers' job would have had to have joined farming every year.

In reality, the farm household replacement ratio in 1950 was 85 per cent (175 000 ÷ 206 000) as calculated in the last row of Table 4.2. The number of farms was abnormally large in 1950, because a large number of urban dwellers and repatriates from overseas territories entered farming during and immediately after the Second World War. In fact, 175 000 inheriting males newly graduated from schools in 1950 were about sufficient to maintain the traditional level of 550 million farms in Japan.

This farm household replacement ratio declined rapidly as non-farm employment opportunities expanded in the process of postwar economic growth. It went down to 56 per cent in 1955, when the period of high economic growth began, and to 14 per cent in 1965. Given such trend, it was natural for the designers of the Agricultural Basic Law to have developed the following expectation:

TABLE 4.2 Changes in the number of new school-leavers in farm households who begin to engage in farming

	1950	*55*	*1965*	*1970*	*75*	*1980*	*85*
School-leavers to engage in farming (000)[a]:							
Total	439	263	60.8	36.9	9.9	7.0	4.2
Male	233	149	41.8	26.1	7.9	5.8	3.6
Inheriting male (A)[b]	175[c]	112[c]	26.8	18.1	6.1	4.0	2.4
Farm households (10 000) (B)	618	604	566	534	495	466	438
Farm household replacement ratio (%) = (A) ÷ ((B)/30))	85	56	14	10	4	3	2

Notes:
a Total of school-leavers from middle school, high school and universities.
b Male school-leavers who are expected to become household heads in future.
c Assumed as 75 per cent of male school-leavers.
Source: S. Namiki, 'Sengo ni okeru Nogyo Jinko no Hoju Mondai (Problems of Replacement of Agricultural Population in the Postwar Period)', *Nogyo Sogo Kenkyhu*, January 1958; JMAFF, *Noka Shugyo Dokochosa* (Survey on Employment Situation of Farm Household Population), *Noringyo Census* (Census of Agriculture and Forestry), and *Nogyo Chosa* (Agricultural Survey).

If the present trend . . . progress[es] further, it will result in a sharp absolute decline in the agricultural labour force. . . . it makes us expect that the number of farms, too, will eventually begin a landslide decline. . . . until today, the problem of surplus labour in agriculture has been considered an invariable exogenous condition in designing agricultural policies and, therefore, the agrarian structure dominated by small-scale peasant farms has been an environment unalterable by policy efforts. Assuming such an agrarian structure as given, the basic policy approach has been to preserve or protect small peasants. But, is it not the situation today that surplus labour is no longer an unalterable condition? A momentum to improve our agrarian structure is now being created through intersectoral occupational mobility (Research Committee on the Basic Problem of Agriculture and Forestry, 1960, pp. 27–8).

In short, policy makers were then aware that the old regime dominated by consideration of the 'poverty problem' and the protection of smallholders was passing away, and that the Japanese economy was reaching a new stage dominated by the 'agricultural adjustment problem'.

4.2.2 The Failing Expectation

With such a perspective, the government promoted policies geared for improvements in the agrarian structure. In 1962, the ceiling of 3 hectares (12 hectares in Hokkaido) on land holdings imposed by the Agricultural Land Law was relaxed. In 1961, the Act of Subsidization for Agricultural Modernization Funds was enacted, under which a part of interest on credit from the agricultural cooperatives to farmers for the purchase of land, livestock and machinery was subsidized by government. In 1962 the Agricultural Structure-Improvement Programme was launched, under which economic development plans at a village level were induced by subsidies to village governments on their investments in land infrastructure and major production and marketing facilities.

Despite these government efforts, the target of 1 million viable farms was far from being achieved by 1970. Instead, the estimated number of viable farms decreased from 521 000 in 1960 to 353 000 in 1970. Moreover, the concentration of agricultural production and resource use in viable farms did not progress, as reflected in changes in the shares of viable farms in agricultural output and inputs in Table 4.3. Indeed, between 1970 and the present time there has been little progress towards a structure dominated by viable farms.

The number of viable farms did not grow in the process of high economic growth, partly because the average income of non-farm employees' households that sets a minimum income for the statistical category of 'viable farms' increased sharply. Most farmers found it difficult to catch up with the rapidly rising income of non-farm employees by expanding the scale of farm operation at a sufficiently rapid rate. For example, a rice farmer was able to earn the minimum income of viable farms from the cultivation of 2.3 hectares of paddy field in 1960. But he needed 3.5 hectares in 1970, 5.5 hectares in 1980 and 5.8 hectares in 1985 to be a viable farmer. Such rapid expansion in farm size for a significant number of farms could have not been possible unless a much larger number of farmers left farming than was anticipated originally by the Basic Law.

TABLE 4.3 Changes in the number and the shares of viable farms in Japanese agriculture

Farm statistics	*1960*	*1970*	*1980*	*1985*
No. of Farms (1000):				
Total	6057	5342	4661	4376
Viable farms	521	353	242	232
Share of viable farms (%):				
Agricultural output	23	25	30	31
Arable land	24	18	19	21
Agricultural labour force	16	19	21	22
Agricultural fixed capital	19	19	21	24

Source: JMAFF, *Noringyo Census* (Census of Agriculture and Forestry); *Nogyo Chosa* (Agricultural Survey); *Showa 61 Nendo Nogyo Hakusho Fuzoku Tokei Hyo* (Statistical Appendix to the 1986 Agricultural White Paper) p. 137.

In reality, such a large-scale rural exodus did not occur. The total number of farms decreased from 1960 to 1970 by only about one-tenth which was much smaller than the original expectation of one-quarter. The decrease in the number of farms altogether for the 25 years from 1960 to 1985 was less than one-third. Even though many farmers and their family members were able to secure more lucrative employment opportunities outside agriculture, they did not migrate to urban areas but remained in rural residence and continued farming in their spare time. Typically, a family head commutes to a factory or a shop in a nearby city and works on the farm only at weekends, while his wife and old parents keep farming in his absence. Because these part-time farm households have increased greatly in number without selling out their land, the concentration of farmland in the hands of viable farmers has been blocked.

The process of structural change in Japanese agriculture for the past 25 years can be appreciated from an analysis of the data in Table 4.4. Corresponding to increases in non-farm employment opportunities, the number of workers mainly engaged in agriculture declined by over 50 per cent from 1960 to 1985. However, this decrease was, to a large extent, compensated for by a sharp increase in the number of workers 'supplementally engaged' in farming while engaged mainly in off-farm jobs.

The increase in 'supplementally engaged' farm workers corresponded to a sharp increase in part-time farm households, especially

TABLE 4.4 Changes in the agricultural labour force, number of farms and arable land area in Japan

Farm statistics	*1960*	*1970*	*1980*	*85*	*1985/ 1960*
Workers engaged in agriculture[a] (0 000):					
Mainly engaged[b]	1454	1025	697	636	0.44
Supplementally engaged[c]	312	522	557	527	1.69
Total	1766	1547	1254	1163	0.66
Farms (0 000)					
Full time	208	83	62	63	0.30
Part-time type I[d]	204	180	100	78	0.38
Part-time type II[e]	194	271	304	297	1.53
Total	606	534	466	438	0.72
(Core farms)[f]		(124)	(103)	(87)	
Arable land (0 000 ha)					
Lowland	338	342	306	295	0.87
Upland	269	238	241	243	0.90
Total	607	580	546	538	0.89
Arable Land area per farm	1.00	1.09	1.17	1.22	1.22
Land utilization rate (%)[g]	133	108	103	104	0.78

Notes:
a Farm household members above 16 years old who engage in farming in any amount of time.
b Engaged in farming more than in other economic activities.
c Engaged in farming less than in other economic activities.
d Farms with farm income larger than off-farm income.
e Farms with farm income smaller than off-farm income.
f Farms with economically active male (16–59 years old) engaged in farming for more than 150 days per year.
g Ratio of total area planted in a year to total arable land area.
Source: JMAFF, *Noringyo Census* (Census of Agriculture and Forestry), *Nogyo Chosa* (Agricultural Survey); *Kochi Oyobi Sakusuke Menseki Tokei* (Statistics of Cultivated Land Areas and Areas Planted in Crops).

of the type II category that earn more than half of their income from non-farm sources. In 1960, full-time, type I and type II part-time farms each occupied one-third of the total number of farms. The share of type II part-time farms increased rapidly, reaching 68 per cent in 1985. Meanwhile, the share of full-time farms decreased to less than 15 per cent.

While the total number of farms decreased very slowly owing to an increase in part-time farms, arable land area decreased slightly due to a conversion of part of it to non-agricultural uses. As a result, the average farm size increased by only 21 per cent over the 25-year period (0.8 per cent per year) – a rate of increase which would require some 90 years merely to double the present 1-hectare size.

Because part-time farmers have held on to their land and continued farming in their spare time by drawing upon supplemental labour from available family members, full-time farmers have found it difficult to expand their operational scale. In 1984, the average land area under cultivation of full-time farm households was only 2.2 hectares, and the average of 'core farms' (those retaining at least 1 male worker from 16 to 59 years old engaged in farming for more than 150 days per year) was only 2.6 hectares.

4.3 DOMINANCE OF PART-TIME FARMING

Why has part-time farming become so pervasive? Of course, the basic reason has been rapid growth in non-farm employment opportunities for farmers within a commutable distance from their residence. While non-farm employment opportunities increased enormously during the regime of high economic growth, the commutable distance expanded to a great extent with progress in motorization and construction of highways.

4.3.1 Population Density and Accessibility to Non-farm Employment

The influence of non-farm employment growth on the increase in part-time farming is especially great in a country like Japan, characterized by high population density on land, because the distance from rural residence to urban employment opportunities is usually relatively short and daily commuting is easy.

The effect of population density and accessibility to non-farm employment is obvious from a comparison between Hokkaido and the rest of Japan. Hokkaido is a northern frontier island settled recently since the Meiji period, where population density is still much lower than in the rest of Japan. For 1960–80, the number of workers mainly engaged in agriculture decreased in Hokkaido by 52 per cent, which was about the same as 55 per cent in the rest of Japan. In Hokkaido, however, the number of farms decreased by 49 per cent as compared with 23 per cent in Japan as a whole. Average farm size increased in Hokkaido by as much as 130 per cent, as compared with only 6 per cent in Japan as a whole. Meanwhile, the share of part-time II farms in the total number of farms decreased from 37 per cent to 26 per cent in Hokkaido, in sharp contrast with a large increase from 32 per cent to 66 per cent in the rest of Japan. The share of viable farms in the total number of farms in Hokkaido reached 31 per cent in 1984, while the national average was only 6 per cent.

The structural design of the Agricultural Basic Law has thus materialized to some extent in Hokkaido. In most other areas in Japan, however, high population density and easy access of rural dwellers to non-farm employment opportunities has resulted in an agrarian structure dominated by part-time instead of viable farms.

4.3.2 Farm Mechanization and Prolonged Life Span

Another major factor underlying the domination of part-time farming was the diffusion of small hand tractors of less than 10 hp, commonly called 'power tillers', during the 15 years of high economic growth after the mid-1950s. The number of power tillers increased at an explosive rate from 89 000 in 1955 to 514 000 in 1960 and to 3 448 000 in 1970, and then declined to 2 752 000 in 1980.

Previously, farm operations in Japan had been largely based on manual labour. Land preparation for rice cultivation especially, had been a very heavy task requiring the labour of young male workers. With the introduction of power tillers, it became possible for female or elderly workers alone to keep on farming; this enabled young to middle-aged males in farm households to engage mainly in non-farm economic activities.

This process was facilitated further by a marked increase in the life span of the farming population. The average life expectancy of the Japanese was only about 50 years in the 1930s and also immediately after the Second World War, but it exceeded 70 years in the 1970s and rose to 75 years for males and 80 years for females in 1985. The increase in the elderly labour force, together with the diffusion of power tillers, made it much easier for rural households to maintain farms without young males working full-time.

Such a process is reflected in increases in the shares of elderly workers in agricultural labour force. The share of females was 48 per cent in 1950, went down to 41 per cent in 1960 and to 39 per cent in 1985, as shown in Table 4.5. From 1960 to 1980 the share of workers above 60 years old in the male labour force increased from 23 to 51

TABLE 4.5 Changes in distribution of gainful agricultural workers by sex and by age in Japan[a]

Sex	Age	% distribution 1960	1970	1980	1985
Male	16–29	77[b]	17	12	8
	30–59		50	45	41
	Above 60	23	33	43	51
	(Above 65)			(31)	(36)
	Total	100	100	100	100
	(Male ratio)	(41)	(39)	(38)	(39)
Female	16–29	86[b]	15	9	6
	30–59		62	59	55
	Above 60	14	23	32	39
	(Above 65)			(20)	(24)
	Total	100	100	100	100
	(Female ratio)	(59)	(61)	(62)	(61)

Notes:
a Gainful agricultural workers are farm household members above 16 years old who engage in farming more than in other economic activities.
b 16–59.
Source: JMAFF, *Noringyo Census* (Census of Agriculture and Forestry); *Nogyo Chosa* (Agricultural Survey).

per cent. In 1985, as many as 36 per cent of male workers in agriculture were older than 65 years, and only 8 per cent of them were younger than 29 years of age.

Thus, in addition to a strong pull of labour from expanding non-farm sectors, small-scale mechanization and prolongation of life expectancy during the 1950s and 1960s worked as a force pushing young male labour out to non-farm economic activities in such a way that part-time farming households became pervasive in rural Japan.

4.3.3 Regulations on Land Tenure

On the institutional side, the regulations of land tenure following the postwar reform represented a major block to structural adjustments. The Agricultural Land Law of 1952 initially restricted ownership of arable land to less than 3 hectares per farm (12 hectares in Hokkaido). Tenancy rights were protected so strongly that it was almost impossible for landlords to evict tenants. In addition, land rent was controlled at a level so low that part-time farmers had little incentive to lease out their holdings. Together, these factors constrained the possibilities of increasing the operational size of farms.

The Land Law was amended in 1962, a year after the enactment of the Agricultural Basic Law, by which the 3-hectare ceiling on land ownership was removed. In this amendment, however, the regulations on tenancy contracts were not relaxed for fear of a revival of landlordism. At that time it was a common expectation of policy makers that the price of agricultural land would come down as many farmers would soon leave farming for urban employment and sell out their land. The authorities therefore expected that the average size of remaining farms could be expanded through transfer of ownership titles if the land ownership ceiling were removed (Kajii, 1985).

In reality, the price of farm land rose sharply corresponding to high economic growth. The land price was raised partly because of government support of agricultural product prices, but more importantly from an expected diversion of farm land to non-agricultural uses – a nationwide construction boom raised farmers' expectation on possible windfall gains from the conversion of their land for factories and highways.

As the result, the price of farm land rose faster than the return to land in its use for agricultural production. As shown in Table 4.6, around the time when the Basic Law was enacted, the rate of return to investment in the purchase of paddy field was about 7 per cent, roughly the same level as prevailed before the Second World War. Within the following two decades it declined below 1 per cent. Such trend indicated that, while the price of farm land had been determined mainly as a present value of expected annual returns for its use in agricultural production until the early 1960s, the expectation of its use for non-agricultural purposes has later become more dominant.

As the price of agricultural land exceeded the present value of an agricultural income stream, it became unprofitable for farmers to enlarge their farms through land purchase. The alternative left for farm-scale expansion was land leasing. In order to activate a land-rental market, the Agricultural Land Law was amended again in 1970, by which rent control was removed and landlords were able to claim a return of their land from tenants upon termination of long-term lease contracts for more than 10 years.

Further, with the amendment of the Agricultural Development Act of 1975, short-term land lease contracts for less than 10 years were legalized where the contracts were agreed jointly by a number of villagers under a village-level farmland utilization programme, as

TABLE 4.6 Farm land prices and rates of return to land purchase in Japan

	Paddy land price (1)[a]	*Farm surplus from paddy land* (2)[a]	*Rate of return to paddy land* (2)/(1)[b]	*Major change in land tenure regulations*
1934	3980	270	6.8	
35	4150	330	8.0	
36	4350	390	8.9	
1952	450[c]	71[c]	15.8	Enactment of Land Laws
55	1160	140	12.0	
1960	1980	140	7.1	
62	2550	161	6.3	Removal of farm-size ceiling
65	3430	194	5.7	
1970	10 220	225	2.2	Freedom to terminate long-term lease contract
72	14 360	245	1.7	
75	28 240	575	2.0	Freedom for short-term lease contract under village-level land-use programme
77	31 600	445	1.4	
79	36 300	315	0.9	
1980	38 280	224	0.6	Freedom for short-term lease contract through mediation of village headman

Notes:
a Yen/hectares.
b (%).
c All statistics after 1952 in 000 yen/hectare.
Source: K. Kajii, 'Nogyo Keiei Kaizen ni Kansuru Shoseisaku no Tenkai' (Development of Policies for Improving Farm Management Improvements), in N. Kanazawa (ed.), *Nogyo Keiei to Seisaku* (Farm Management and Policies) (Chikyunsha, 1985) p. 65. The 1980 data are supplemented.

they are exempt from the application of the Land Law. Finally, by the Farmland Utilization Promotion Act of 1980 short-term contracts agreed upon through the mediation of village headmen also became exempt from the application of the Land Law and farmland under such contracts is to be returned automatically to landlords upon termination of the contract period.

TABLE 4.7 Distribution of farms that experienced net increase in operational land-holding in Japan

	Total no. (000)	*Distribution (%)*			
			Net increase through		
		Total	*Purchase*	*Lease*	*New land opening and other*
1961	359	100	37	18	45
65	271	100	34	23	43
68	194	100	28	22	50
1971	150	100	34	28	38
74	185	100	24	29	46
77	172	100	16	40	44
79	202	100	13	47	40
1981	186	100	16	56	28
83	179	100	16	45	39
84	159	100	16	46	38

Source: JMAFF, *Nogyo Chosa* (Agricultural Survey).

With such deregulation of the land market, land leasing has become a major instrument for farmers to enlarge the scale of their operation. As shown in Table 4.7, land lease had been a less important factor than land purchase in accounting for land transfers until the 1970 amendment of the Land Law. Since then, it has become a much more important factor.

Even though a land-rental market has thus been activated, the scope of farm-size expansion by means of land lease has yet been very limited. The 1985 Census of Agriculture shows that the share of leased-in land in the total farming area was only 7.5 per cent which was very much lower than the 20–50 per cent in Western Europe and North America. Regulations on land tenure, even though relaxed gradually, have been a major institutional impediment for the development of viable farms in Japan.

The problem is that this barrier cannot be removed simply by changing laws. So long as the memory of land confiscation in postwar land reform is alive in the memory of rural people, a fear continues to prevail that land, once let, is lost forever. Non-farmers who inherit land thus tend to hold on to it and farm it part-time rather than leasing or selling it out, because their attachment to the land has been buttressed by a number of factors, among them an ancestral identification with the land, a desire to maintain the farm as a productive

TABLE 4.8 Shares of various categories of farms in Japanese agriculture, 1984

	No. of farms	*Arable land area*	*Agricultural output value %*					
			Total	*Rice*	*Poultry*	*Pig*	*Dairy*	*Greenhouse vegetables*
Full-time	15	27	29	16	27	39	51	36
Part-time I[a]	15	30	40	32	60	49	40	50
Part-time II[b]	70	43	31	52	13	12	10	15
Total	**100**	**100**	**100**	**100**	**100**	**100**	**100**	**100**
Core farm[c]	20	47	59	32	72	80	90	83
Other	80	53	41	68	28	20	10	17
Total	**100**	**100**	**100**	**100**	**100**	**100**	**100**	**100**
Viable farm[d]	5	21	31	11	56	50	63	45
Other	95	79	69	89	44	50	37	55
Total	**100**	**100**	**100**	**100**	**100**	**100**	**100**	**100**

Notes:
a Farms with farm income larger than off-farm income.
b Farms with farm income less than off-farm income.
c Farms with economically active male (16–59 years old) engaged in farming for more than 150 days per year.
d Farms with average agricultural income per full-time worker equivalent to the average income of a non-farm employee.
Source: JMAFF, *Showa 61 Nendo Nogyo Hahusho Fuzoku Tokei Hyo* (Statistical Appendix to the 1986 Agricultural White Paper), pp. 136–7.

post-retirement asset, and the expectation of higher land prices in the future.

4.4 CONSEQUENCES OF THE DOMINANCE OF PART-TIME FARMING

The spread of part-time farming has thus impeded attempts by full-time and core farmers to enlarge operational land holdings. Their efforts to expand farm production and income have therefore concentrated on enterprises less dependent on land. As shown in Table 4.8, the shares of full-time and core farms in farm output are high in land-saving and capital-intensive enterprises such as poultry, pigs and greenhouse vegetables. Their share in dairy production is also high, despite its land-using nature, because dairy production is concentrated mostly in relatively land-abundant Hokkaido.

In contrast, the shares of part-time II farms and non-core farms with no male worker between 16 and 59 years old mainly engaged in farming are high in land-intensive rice production. Part-time and elderly farmers tend to concentrate on rice farming because it is a very stable crop offering a high return on only intermittent labour without much managerial effort. Because rice marketing is carried out exclusively by the government, rice farmers are guaranteed a high price and can easily sell their harvest through agricultural cooperatives, the sole agents of government rice marketing. In addition, agricultural research and extension services have traditionally concentrated on the rice crop to the extent that rice cultivation has become highly standardized and there is little difference in productivity between part-time and full-time farmers, and between young and elderly farmers. The fact that the production of Japan's staple crop has been geared to part-time farming in this way is thus a major factor encouraging part-time farming and impeding the concentration of farm land in the hands of farmers wishing to expand their operational land holdings to a viable scale.

The spread of part-time farming has been a major factor preventing the realization of the structural design of the Agricultural Basic Law. It should be pointed out, however, that the increase in off-farm earnings has made it possible to achieve the prime policy goal of the Basic Law – i.e. income equalization between farm and non-farm families. As seen in Table 4.9, in 1960 average farm household income was well behind urban wage earners' income, by 11 per cent on a per household basis and 32 per cent on a *per capita* basis. Yet this disparity narrowed during the process of rapid economic growth until *per capita* farm household income exceeded *per capita* urban wage-earner income in 1975. The main force behind this rapid increase in farm household income has been the off-farm earnings. Non-agricultural income rose from approximately 50 per cent of all farm household income in 1960 to 85 per cent in 1985. The importance of this non-agricultural income is also evident from the ironic fact that part-time II farmers whose average scale of operation is the smallest have the highest incomes, and full-time farmers whose farm size is the largest, the lowest.

It may appear anomalous in Table 4.9 that not only off-farm income but also farm income per household is absolutely smaller in full-time than in part-time I households. This anomaly is explained by the fact that the category of full-time farm households include those consisting of elderly members alone who are unable to work hard on the farm, not to speak of off-farm employment. In fact, the numbers of full-time farm households holding no member below 65 years old increased especially sharply in recent years, from 196 000 in absolute numbers (or 32 per cent of the total number of full-time farms) in 1980 to 237 000 or 38 per cent. This increase in 'old-aged' full-time farms underlay a stability in the total number of full-time farms during the recent inter-census period from 1980 to 1985 (Table 4.4, p. 82). A slight increase in the number of full-time farms at the expense of part-time II farms recorded for the 1980–5 period for the first time since the beginning of high economic growth reflects a retirement process of farm household members from off-farm employment. To the extent that those retiring people hold on to land and farm it in a leisurely fashion, the route continues to be blocked for core farms to accumulate land and rise to a viable unit.

The spread of part-time farming has had important political implications. It has contributed greatly to the strengthening of the political power of the farm bloc organized by *NOKYO* (agricultural cooperative organization). The spread of part-time farming prevented a decline in farm population and the number of votes in support of the farm bloc. Concurrently, adjustments in electoral districts have lagged significantly behind changes in the interregional distribution of population. As a result, rural votes have had a disproportionately higher weight in representation to the National Diet than urban votes (one to three in extreme cases). These factors have combined to prevent a decline in the political influence of the farm bloc despite growing concentration of population in urban areas.

Part-time farmers (especially of the type II category) are the economic and political basis of *NOKYO* activities. They are obedient and faithful customers to *NOKYO* in the sale of farm inputs and the marketing of their produce. While large full-time farmers often make hard bargains and try to obtain better deals through a deliberate choice among cooperatives and private traders, part-time II farmers tend to leave business to a nearby village cooperative with no hard bargaining, because their time commands a high opportunity cost and their product sale and input purchase are individually too small to be worth expending a large effort.

Part-time II farmers are also silent but obedient followers of *NOKYO* politics. They are high-cost producers and are not inclined to spend time and energy in raising farm productivity. Naturally they support and vote for agricultural protection policies, even though the benefits they receive from such policies are very minor relative to

TABLE 4.9 Comparisons of income levels between farm household and urban worker households in Japan

	Farm household income (000 yen)				*Urban worker household Income (000 yen)*		*Relative income (%)*	
	Per household							
	Farm	*Off-farm*[a]	*Total (A)*	*Per household member (B)*	*Per household (C)*	*Per household member (D)*	*Per household (A/C)*	*Per household member (B/D)*
1955	256 (72)[b]	102 (28)	358 (100)	57	350	74	102	77
1960	225 (50)	224 (50)	449 (100)	78	502	115	89	68
65	365 (44)	470 (56)	835 (100)	157	797	194	105	81
1970	508 (32)	1084 (68)	1592 (100)	326	1390	358	115	91

75	1146 (29)	2815 (71)	3961 (100)	867	2897	760	137	114
1980	952 (17)	4651 (83)	5603 (100)	1273	4254	1111	132	115
1985	1065 (15)	5860 (85)	6926 (100)	1596	5388	1422	128	112
1985								
Full-time	2520 (56)	1969 (44)	4489 (100)	1207	5388	1422	83	85
Part-time I	4124 (56)	3275 (44)	7399 (100)	1465	5388	1422	137	103
Part-time II	495 (7)	6971 (93)	7466 (100)	1682	5388	1422	139	118

Notes:

a Off-farm income includes private grants and government transfer payments.

b Percentage shares are shown in parentheses.

Source: JMAFF, *Noka Keizai Chosa* (Farm Household Economy Survey). Japan Prime Minister's Office of General Administration, *Kakei Chosa* (Household Survey).

off-farm incomes. The interests of part-time II farmers and *NOKYO* especially coincide in the preservation of direct government control on rice price and marketing. The rice control system guarantees a high stable price and easy marketing for part-time II farmers while it guarantees a large commission to *NOKYO* as a sole agent in collecting rice for the Food Agency.

In general, part-time II farmers in Japan do not show a strong propensity to be 'free riders' in the *NOKYO* movements, despite their large number and the negligible impact of individual participation on the outcome. Part of the explanation might be sought in social compulsion based on intensive social interactions (Becker, 1974) in traditional village communities. Japanese village communities are known to be 'tightly structured', even compared with other Asian communities, where people have a strong inclination to conform to group norms and decisions (Embree, 1950). Another reason might be the provision by *NOKYO* of 'by-products' (Olson, 1965): in addition to agricultural marketing and political lobbying (main products), *NOKYO* supplies a number of 'by-product' services such as mutual insurance and credit, which are attractive even to non-farmer residents in rural communities. In fact, that explains why *NOKYO* is able to maintain 2 million associate members who are non-farmers, in addition to 5 million farmer members.

Part-time farmers and *NOKYO* have thus reinforced each other to dominate the rural economy and politics of Japan. As a result, it has been more and more difficult to achieve the major structural adjustments in agriculture needed to escape from the stalemate of agricultural protection policies.

5 New Prospects for Structural Adjustment

The Japanese government's efforts since the 1961 Agricultural Basic Law to foster efficient agriculture by concentrating land and other agricultural resources in a small number of large-scale 'viable farm units' have not been successful so far. Mainly this is because of the nationwide spread of type II part-time farming. However, there are now signs that conditions are ripe for achieving the structural adjustments envisaged at the time of the Basic Law.

With a series of amendments to the Land Laws (1962, 1970 and 1980) and under the Farm Land Utilization Promotion Law in 1980 (by which contracts for short-term land leasing with the mediation of village headmen are exempted from the regulations of the Land Laws) institutional barriers for farm-size expansion through land leasing have largely been removed. Concurrently, changes in agricultural technology and labour markets have produced conditions favourable for large-scale viable farms. This new prospect for structural adjustment is investigated in this chapter.

5.1 FARM MECHANIZATION AND SCALE ECONOMIES

As explained in Chapter 4, small-scale mechanization (represented by the spread of power tillers in the 1950s and 1960s) facilitated substitution of elderly and/or female labour for young male labour and thereby contributed to the prevalence of part-time farming. However, it appears that a new surge of mechanization which began in the late 1960s has been undermining the dominance of small-scale part-time farming.

As shown in Figure 5.1, farm mechanization in Japan progressed in three stages. The first stage was the diffusion of static machines such as threshers and water-lifting pumps that began before the Second

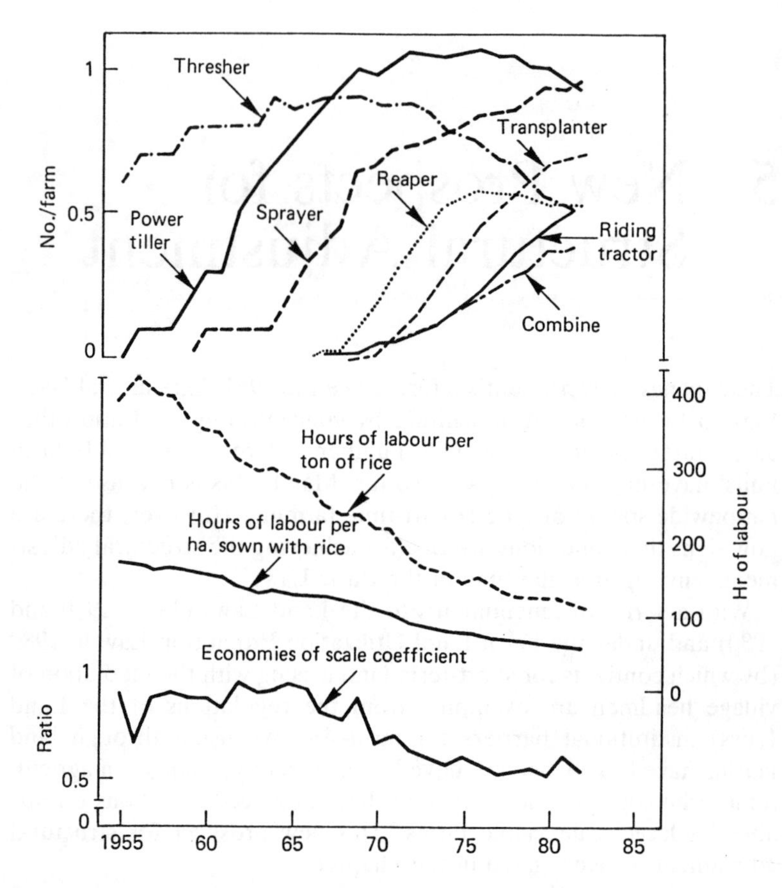

FIGURE 5.1 Agricultural mechanization and economies of scale in Japanese agriculture

Note: Economies of scale coefficient =

$$\frac{\text{Rice Production cost in large farms (larger than 3 hectares)}}{\text{Rice production cost in small farms (smaller than 0.3 hectares).}}$$

Source: JMAFF, *Nogyo Chosa* (Agricultural Survey); *Kome Seisanhi Chosa* (Rice Production Cost Survey)

World War. The second stage was mechanization with small-scale mobile machines, such as power tillers and sprayers, during the 1960s. This was only a partial mechanization, leaving several tasks (such as transplanting and harvesting) to manual labour. The third stage that has been taking place since the late 1960s is marked by large-scale mechanization with riding tractors, combines and rice transplanters. In this stage the whole process of rice farming has become mechanized.

In labour-saving terms, these three stages of mechanization seems to be continuous, judging from continuous decreases in labour input both per ton of rice and per hectare of rice area as shown in the middle section of Figure 5.1. However, as indicated in the bottom graph in Figure 5.1, the third stage differs from the second in that in the former the average cost of rice production of large farms (above 3 hectares) was nearly one-half that of small farms (below 0.3 hectares), while in the latter there was no significant difference. This contrast shows that the third-stage mechanization is characterized by increasing returns to scale, while small-scale mechanization in the second stage was neutral with respect to scale.

Emergence of scale economies has not been unique to rice, but has been common to other crops and livestock production in more recent years as wage rates have risen to a level comparable with those of North America and Western Europe and the need for saving labour has become greater. The estimation of an aggregate agricultural production function in Japan based on interregional cross-section data shows that the sum of production elasticities was close to 1 before 1965, but that it became significantly larger than 1 after that date (Kuroda and Yoshida, 1981). Such results for Japan are consistent with the results of estimation of an aggregate production function based on intercountry cross-section data that show significant scale economies in agricultural production in high-income countries, in contrast with scale neutrality in low-income countries (Hayami and Ruttan, 1985, Chapter 6; Kawagoe, Hayami and Ruttan, 1985).

Farm mechanization in the third stage has been characterized by increasing returns because it needs a large farm unit to operate large-scale machinery efficiently. The effect of indivisible machinery capital on decreasing average cost to farm size is reflected by the data in Table 5.1. In the second-stage mechanization until 1965, relative to small farms, large farms spent less on labour but more on machinery and power per unit of rice output. In the third stage the costs of both labour and machinery have been lower for large as compared with small farms. Such contrast indicates that the scale advantage of utilizing machinery capital did not exist so long as small machinery such as power tillers were used commonly, but scale advantages have emerged corresponding to the introduction of large-scale machinery such as riding tractors and combines.

TABLE 5.1 Comparison of rice production costs in Japan by farm size, 1955–84

Cost item	Farm size (ha)	Index of production cost per kg (Below 0.3 ha = 100)						
		1955	*1960*	*65*	*1970*	*75*	*1980*	*84*
Total costs	Below 0.3	100	100	100	100	100	100	100
	0.0–0.5	103	106	102	95	93	93	90
	1.0–1.5	98	96	88	78	75	70	67
	2.0–3.0	88	92	82	67	61	57	54
	Above 3.0	90	87	93	69	60	51	51
Labour costs	Below 0.3	100	100	100	100	100	100	100
	0.3–0.5	100	106	100	98	93	92	92
	1.0–1.5	91	91	87	79	72	66	62
	2.0–3.0	74	83	80	63	52	50	49
	Above 3.0	73	78	90	68	50	42	43
Machinery and power costs	Below 0.3	100	100	100	100	100	100	100
	0.3–0.5	138	135	134	104	102	103	91
	1.0–1.5	156	146	127	89	90	81	72
	2.0–3.0	143	137	108	72	82	64	54
	Above 3.0	144	121	123	69	76	55	51

Source: JMAFF, *Kome Seisanhi Chosa* (Rice Production Cost Survey) various issues.

However, the observed scale economies do not at all imply that the optimum farm size is so large as to require a large hierarchical organization beyond the scale of a family farm. Current farm mechanization in Japan with tractors of 20 to 40 hp may be more properly called 'medium-scale' rather than 'large-scale'. An optimum farm size for the most efficient utilization of such medium-scale machinery (at a minimum average cost) would be no larger than the average farm size in North Western Europe and definitely smaller than in North America and Australasia. However, there is no doubt that the optimum size is far larger than Japan's average of around 1 hectare.

Several efforts have been made to utilize large machines efficiently in small farms, such as the organization of machinery utilization cooperatives and contract farm operations by machine owners. Yet such efforts have failed to overcome inefficiency in the use of 'medium-scale' machinery in very small farms in Japan, as is evident from the data in Table 5.1. In short, the increasing returns to scale that characterize Japanese agriculture today stem from a lag in farm-size adjustment to a new technology introduced for the sake of greater labour saving, as the wage rate in Japan has risen to a level comparable with those of other advanced industrial economies.

5.2 POLARIZATION IN PROGRESS

Given the emergence of scale economies, it is expected that large farms will increase their operational size by acquiring land from smallholders. Are the economies of scale in Japanese agriculture sufficiently strong now to warrant progress in polarization in the form of farm size expansion in larger farms and size reduction in smaller farms?

5.2.1 A Condition for Large-scale Tenant Farming

The price of agricultural land in Japan today is far above a level that would justify the purchase of land for the purpose of agricultural production (Table 4.6, p. 87). Therefore, if a farmer wants to enlarge his operational holding, he must rely on the land-rental market. A question is whether the economic advantage of large over small farms is now so large that large farmers are able to pay a sufficiently high rent to induce small farmers to stop farming and rent out their land.

For a large full-time farmer who intends to rise to a viable unit, the rate of rent he is willing to pay would be the value of total agricultural output per hectare minus the cost of all non-land inputs including imputed costs of unpaid family labour, managerial ability and own farm capital, if the labour and capital markets are working efficiently. In practice, however, it is difficult to divert farm-specific managerial ability and capital such as farm buildings to non-farm uses. In the short run, at least, the opportunity cost of such farm-specific capital (both human and material) may be close to zero. The maximum amount of rent per hectare he is willing to pay for would therefore be a 'surplus' from farming, defined as a residual of output after subtracting the costs of purchased current inputs, hired labour and capital services and the imputed family labour wages.

Unlike large full-time farms, in which farming is carried out mainly by young to middle-aged operators whose labour commands a high opportunity cost, small part-time farms depend for their operations mainly on the labour of wives and old parents with little opportunity cost. They would not stop farming and rent out their land unless the whole income from farming consisted of economic returns, not only to land and farm-specific capital but also compensation for unpaid family labour if the marginal utility of labour by family members is negligible.

In such a situation a necessary condition for the development of large-scale tenant farming may be specified as:

Surplus of large farms
$\geqq$ income of small farms – utility of family labour used in small farms

where 'income' is defined as the value of total farm output minus the costs of current inputs, hired labour and capital services per hectare, and 'surplus' is 'income' per hectare minus imputed family wages.

It is difficult to estimate the utility of family labour. However, if the above condition is met under the assumption of zero labour utility, we are safe to infer that scale economies operating in Japan are sufficient to assure the condition for development of large-scale tenant farming.

The time-series comparison between the surplus of large rice farms with more than 3 hectares of paddy fields and the income of small rice farms (below 0.3 hectare) on a per-hectare basis is shown in Figure 5.2. In the case of Japan as a whole, shown in the top graph, before 1965 the surplus of large farms had not been so much larger than that of small farms. Until this time, too, the large farm surplus was far lower than the income of small farms. This indicates that the condition for the development of large-scale tenant farming had not been attained. The difference in income per hectare between large and small farms began to diverge in the latter half of the 1960s. The gap between the surplus of large farms and the income of small farms became correspondingly, narrower and was completely closed by the mid-1970s. Such a process, when compared with that in Figure 5.1, clearly shows that the economies of scale created from large-scale mechanization had prepared the condition for the development of large-scale tenant farming.

The periods by which this condition was established were different among Japan's regions. Among ten major agricultural regions according to the JMAFF classification, Hokuriku and Tohoku represent two extreme cases. Whereas the condition was already met in the 1970s for the Hokuriku region (middle graph in Figure 5.2), it was

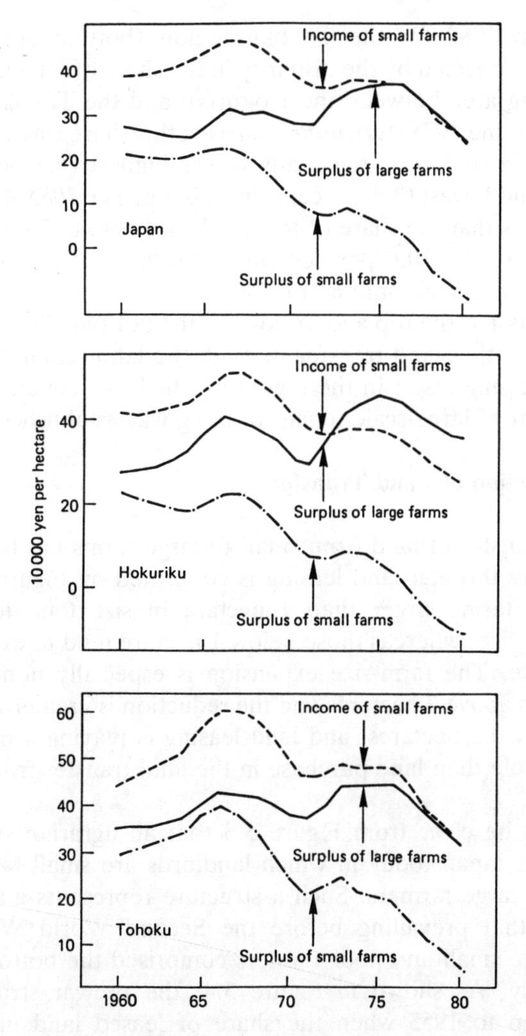

FIGURE 5.2 Comparison between average surplus from rice production per hectare of large farms (larger than 3 hectares) and average income from rice production per hectare of small farms (smaller than 3 hectares), deflated by the rural consumer price index (five-year moving averages)

Note: Income = output value − (material input cost + wages for hired labour)
Surplus = income − wages for family labour

Source: JMAFF, *Kome Seisanhi Chosa* (Rice Production Cost Survey); *Hoson Bukka Chingin Tokei* (Statistics of Rural Prices and Wage Rates)

met only in 1980 for the Tohoku region (bottom graph). Such difference is reflected by the disparity in the share of tenanted land in total farming area between the Hokuriku and the Tohoku regions. According to the *1978 Agricultural Survey*, this share was 4.8 per cent for the whole of Japan; regionally it was highest (6.8 per cent) in Hokuriku and lowest (3.4 per cent) in Tohoku. The *1985 Agricultural Census* shows that the share of tenanted land rose to 7.0 per cent in Japan as a whole, 10.6 per cent in Hokuriku and 5.1 per cent in Tohoku; the regional rankings of 1985 remained the same as of 1978 with Hokuriku at the top and Tohoku at the bottom. These observations indicate that land transfer through the land-rental market has been progressing faster in the region in which the condition for the development of large-scale tenant farming was established earlier.

5.2.2 Direction of Land Transfer

That the transfer of land from small to large farms has been taking place mainly through land leasing is confirmed by Figure 5.3. This shows that farms larger than 1 hectare in size tend to increase operational size, whereas those below 1 hectare tend to experience a net decrease. The farm-size expansion is especially noticeable for larger farms above 2 hectares and the reduction is greater for smaller farms below 0.5 hectares, and land leasing is playing a much more important role than land purchase in the land transfer from small to large farms.

It should be clear from Figure 5.3 that an agrarian structure is emerging in Japan today in which landlords are small farmers and tenants are large farmers. Such a structure represents a sharp contrast with that prevailing before the Second World War, where tenants were small and poor farmers comprised the bottom class in rural society. As shown in Figure 5.4, the prewar structure still prevailed up to 1955 when the share of leased land in the total operational holdings of farms was inversely correlated with farm size. Later the inverse relation became weaker; a positive correlation began to appear in 1970, and has become progressively stronger since then. These changes imply that land leasing from small to large farms has been a continuous process, increasing especially rapidly from around 1970.

Comparing Figures 5.1 through 5.4 indicates clearly that the scale economies created from large-scale mechanization since the late 1960s have prepared the condition for the development of large-scale

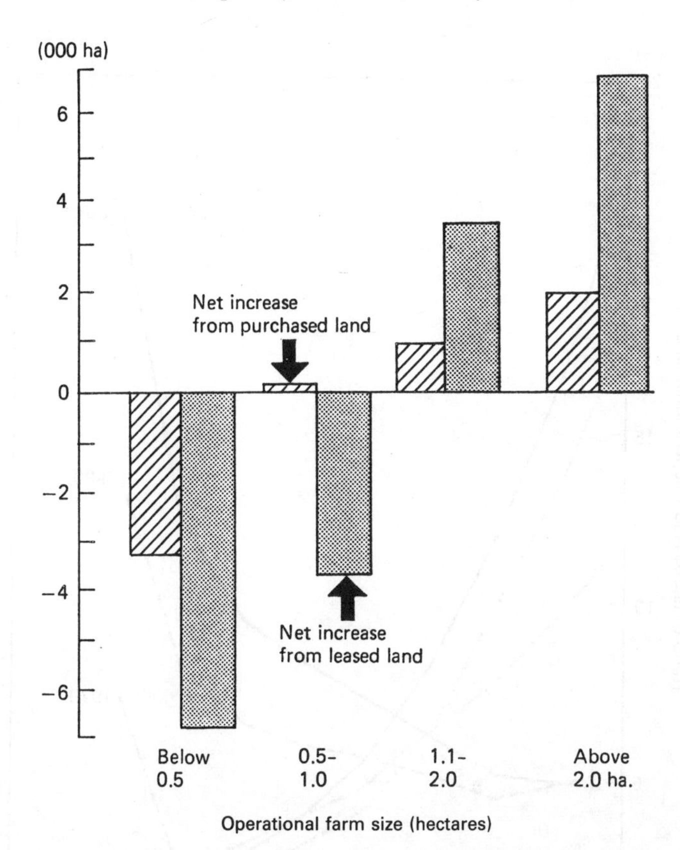

FIGURE 5.3 Changes in operational farm size made through purchases and leasing in Japan (excluding Hokkaido) by farm size class (1982)

Source: JMAFF, *Showa 58 Nendo Nogyo Hakusho* (1983) (Agricultural White Paper) p. 170

tenant farming. Correspondingly, polarization has been in progress: there has been farm-size expansion by large farmers and a reduction of small farmers through land leasing from the latter to the former. Technological conditions for structural changes leading to the dominance of large-scale viable farms have now come to fruition.

It should be pointed out that there are various routes to reach the target of a viable farm unit. In addition to the direct route through land-leasing contracts between large and small farms, a route is open via contracts for custom work for individual farm tasks and cooperative farming.

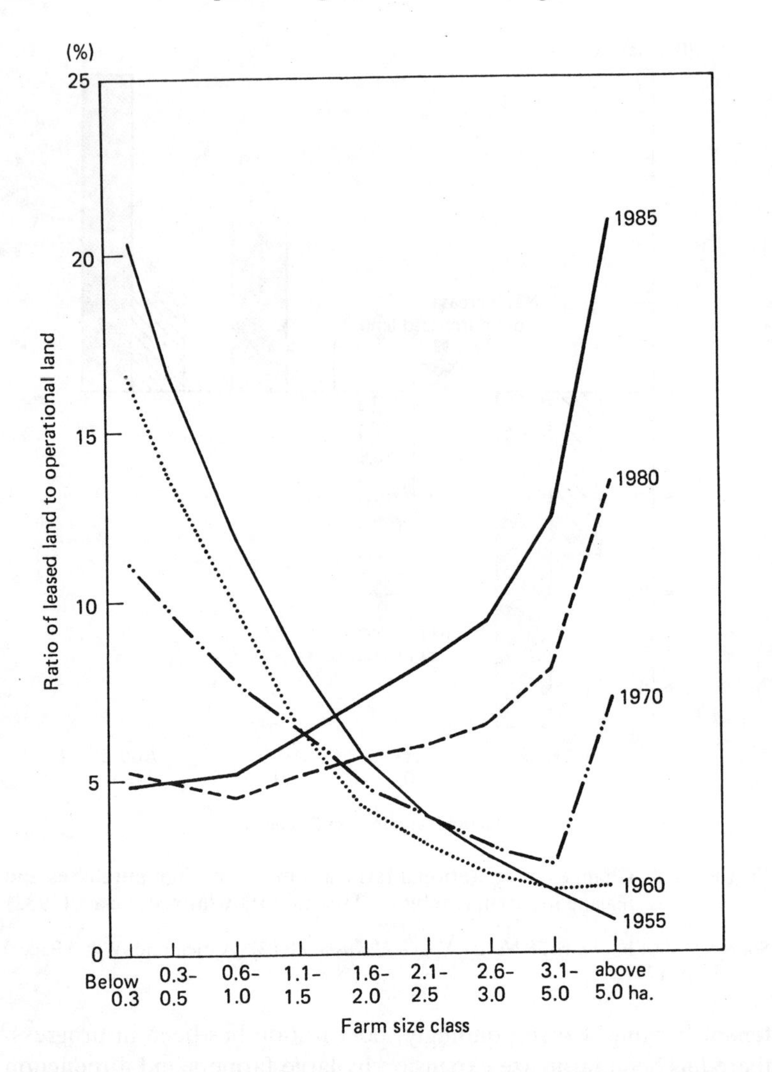

FIGURE 5.4 Ratios of leased land to total operational land holdings by farm size class in Japan (1955–85)

Source: JMAFF, *Noringyo Census* (Census of Agriculture and Forestry); *Nogyo Chosa* (Agricultural Survey)

It has been common for small-scale part-time farms to contract out certain tasks such as land preparation and harvesting to be conducted by large farmers who owned tractors and combines. In many cases, contracts for specific tasks shifted gradually to cover all tasks, ending up in a contract similar to land-leasing. Also, it had been common to establish a production cooperative at a village level to which villagers consigned the use of land under the operation of a few full-time workers; the residual of output after deducting production costs including operators' wages was usually divided among villagers in proportion to the areas of land they contribute.

In both the individual contract of custom work and cooperative farming, as the relation between operators and suppliers of land becomes more enduring and stable, it is natural to shift to a land-leasing contract, either formally or *de facto*, under which operators have greater incentive to exert their entrepreneurial ability while shouldering the risks. In this way, both the private contract of custom farm work and cooperative farming can be considered a step towards the establishment of large-scale tenant farming.

5.3 LABOUR MARKET PROSPECTS

Size expansion of large farms through land-leasing, as observed in the previous section, has been promoted not only by scale economies stemming from large-scale mechanization but also by the conditions of the labour market. For the past several decades the development of viable farms has been blocked, partly because part-time farming has been supported by prolongation of the life span of the rural population.

5.3.1 Conditions of Structural Adjustment

The increasing reliance of part-time farming on elderly labour is now reaching a limit. In the 1980s the share of farms with operators older than 60 years approached 30 per cent, and half of them had no son who expects to succeed in farming. It is reasonable to assume that these elderly operators will have to retire relatively soon, leaving their farms to neighbours.

Meanwhile, the young male labour force engaged mainly in agriculture has continued to decrease. As shown in Table 5.2, the number of 'male core farm workers' who were older than 15 years and mainly engaged in work in their own farms declined from a level of 5 million in 1960 to 1.9 million in 1985. Today, less than 30 per cent of these 1.9 million male core workers are below 50 years old, and

TABLE 5.2 Changes in the number of male core farm workers aged over 15 and mainly engaged in work on their own farms

(000 workers)

Age	*1960*	*1970*	*1980*	*1985*
16–19	268	57	8	4
20–29	1095	302	151	90
30–39	1141	565	212	197
40–49	909	735	395	245
50–59	1009	644	554	526
Above 60	1043	918	716	807
Total	**5465**	**3222**	**2036**	**1870**

Source: JMAFF, *Noringyo Census* (Census of Agriculture and Forestry).

only about 15 per cent below 40 years. The number of boys in farm families who leave school and begin farming with the expectation of becoming farm operators upon the retirement of their fathers is only about 3000 a year. The farm household replacement ratio is only about 2 per cent (Table 4.2, p. 79), implying that the number of farms will decrease from its present 4.3 million to only about 80 000 within a working-life span of one generation, unless the boys who migrate out from farm to urban occupations at the time of school leaving return to farming in later years. In fact, the number of male core farm workers below 30 years old is now already less than 100 000 altogether. All these statistics suggest the likelihood that a major decline in the number of farms will begin within a next decade or so.

Moreover, non-farm employment opportunities for people remaining in farms have become more stable and lucrative. As shown in Table 5.3, in 1960 less than half the part-time farm workers were able to earn salaries from permanent off-farm employment opportunities such as factories and offices, and the rest had to rely on unstable and less lucrative employment such as casual work in construction, self-employed petty trades and cottage industries. Today, 64 per cent of them are engaged in permanent employment; this percentage is markedly higher for younger generations, reaching almost 90 per cent for those below 20 years old. It seems reasonable to expect that both the disutility of farm work *and* its opportunity cost will become progressively larger for younger generations with higher and more stable incomes from non-farm sources. If so, when they become household heads, they will be more tempted to stop farming and rent out their land than their parents today, whose work ethic

TABLE 5.3 Distribution of part-time farm workers in Japan by non-farm economic activity in which engaged

	No. of part-time workers[a] *(0 000)*	*Distribution (%)*			
		Total	*Permanent employment*	*Temporary employment*	*Self-employment*
1960	637 (36)[b]	100	44	19	37
1970	858 (55)	100	49	32	19
1980	817 (65)	100	61	22	17
1985:	745 (64)	100	64	20	16
Age					
16–19	12	100	90	7	3
20–29	163	100	88	7	5
30–39	192	100	72	14	14
40–49	150	100	59	22	19
50–59	162	100	48	32	20
Above 60	66	100	28	34	38

Notes:
a Part-time workers are defined as those employed in non-farm jobs for more than 30 days per year or engaged in self-employed non-farm economic activities that earn more than a certain amount of revenue (100 000 yen per year in 1986).
b In parentheses are percentages of part-time workers in the total number of workers engaged in agriculture.

Source: JMAFF, *Noringyo Census* (Census of Agriculture and Forestry); *Nogyo Chosa* (Agricultural Survey).

was nurtured in poverty and a subsistence crisis common to farm households before the beginning of high economic growth.

Two decades after the enactment of the 1961 Agricultural Basic Law, conditions are thus now becoming ripe for major structural adjustment in Japanese agriculture as was originally envisaged by the Law. In fact, Shokichi Namiki (who provided a theoretical basis for the Basic Law) has recently made the following observation:

> Is it not likely that a major change will occur in the systems of rice farming when the generation born in the first decade of the *Showa era* [1926–35] retire? . . . My forecast that a landslide decline in the number of farm boys who intended to succeed to farming would be followed by a landslide decline in the number of farms did not materialize, because life expectancy became longer and small-scale mechanization had no scale economies. However, the effects of these two factors are now about to end. After a considerable time lag beyond my original expectation, a condition is now being gradually established for viable farms to increase their share in the production of land-using crops, such as rice (Namiki, 1981, p. 9).

5.3.2 Barriers to Structural Adjustment

Although the basic condition is being satisfied for the polarization of farm units, there are still great difficulties in achieving the structural adjustment at sufficient speed to ensure that agricultural productivity growth rises fast enough to prevent a decline in agriculture's comparative advantage.

First, the traditional pattern of land holdings presents a major barrier. In Japan, not only average farm size is small but a farmer cultivates typically a large number of small plots scattered around non-contiguously. When a farmer tries to accumulate land, there is usually no way but to acquire those small plots one by one. Until land accumulation reaches a stage where these plots can be consolidated into a decent contiguous unit, it would be difficult fully to exploit the merit of increasing returns from the efficient use of large-scale machinery. Meanwhile, the advantage of large over small farms would not rise to such a level that large farmers could pay a sufficiently high rent to induce small farmers to lease out their holdings. A vicious circle thus tends to operate between the difficulty of consolidating small scattered plots and the relative inefficiency of large farms which consist of numerous parcels of land.

Second, the flow back of elderly labour from the non-farm to the farm sector that began with the end of high economic growth in the mid-1970s has been becoming a serious impediment to structural adjustment. As shown in the upper graph of Figure 5.5, the number of workers who migrated out from farm to non-farm occupations had been very large for a decade and half of high economic growth until the first Oil Crisis of 1973. For that period, not only relatively young

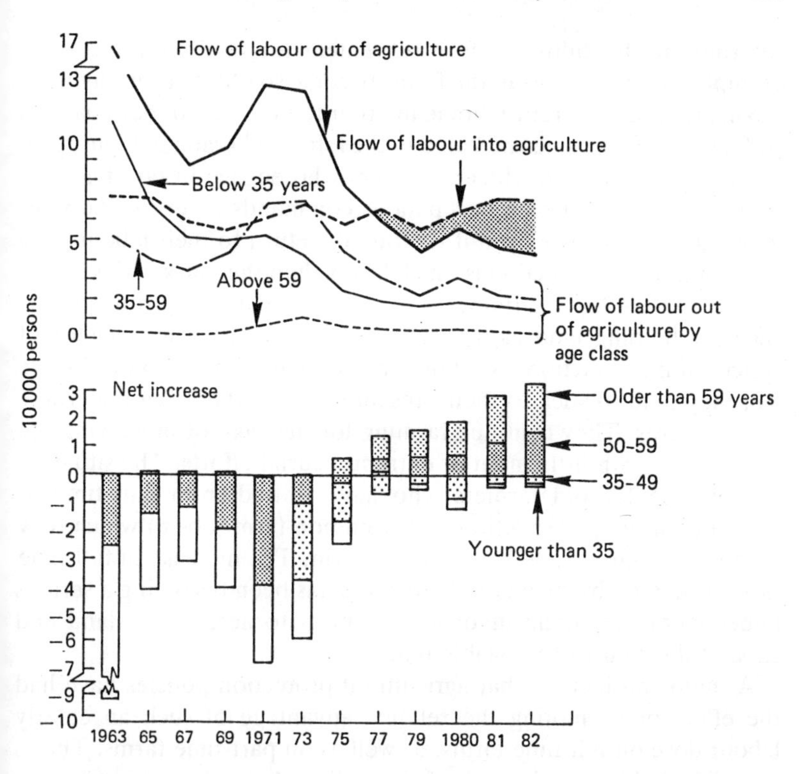

FIGURE 5.5 Flow of the male labour force from agriculture to other sectors and its flow from other sectors into agriculture

Source: JMAFF, *Noringyo Census* (Census of Agriculture and Forestry); *Nogyo Chosa* (Agricultural Survey)

workers below 35 years old but also middle- to old-aged workers between 35 and 60 years had moved out from agriculture.

Labour outmigration dropped sharply in the recession following 1973, and since then there has been a net inflow into agriculture of workers hitherto engaged in non-farm economic activities – although a majority of the sons and daughters of farm families entering the labour market on leaving school has continued to be absorbed by the urban sector (Table 4.2, p. 79). The lower graph in Figure 5.5 indicates that the net labour inflow for recent years has consisted mainly of elderly workers who migrated out to urban jobs earlier and have been reaching the age of retirement. Some of them have been returning to agriculture before retirement for family reasons – for example, where parents in the farms become so old that sons living in urban areas have to return home in order to take care of both parents and farm properties, as indicated in the left-hand graph in Figure 5.6.

The labour of those elderly workers who engage in farming after coming back from urban occupations commands a very low opportunity cost. Therefore even if the rate of return to their labour used for agricultural production is much lower than the market wage rate, they would not leave farming. In general, they have a strong preference for holding land as a secure asset rather than as a factor of agricultural production. As shown in the right-hand graph in Figure 5.6, they tend to stick to their present status, rather than undertake new ventures. They continue farming for the sake of preserving the farm assets, with little input of entrepreneurial efforts. The situation is the same for old farmers who have shifted from part-time to full-time farming after retirement from non-farm jobs to which they used to commute while living on the farm. The increase in full-time farms operated by such elderly workers has been becoming a serious impediment for young innovative farmers to accumulate land and expand their farms to a viable unit.

A major problem is that agricultural protection policies have had the effect of promoting the relative advantage of such an elderly labour dose on full-time farms, as well as on part-time farms. This is especially evident in the case of rice policy. As explained in Chapter 4, government controls on rice prices and marketing reduce the premium on farm operators' entrepreneurial and managerial abilities, and thereby reduce differences in profitability between farms managed by young innovative operators working full-time and those run by part-time farmers or elderly farmers retired from urban occupations. Moreover, acreage control is forced on all producers in equal proportion to their land holdings which prevents the full exploitation of scale economies by large full-time farmers. In addition, the very low rate of taxation on farm land increases the advantage of farmers holding land for the purpose of preserving land as a secure asset relative to those trying to use it efficiently for agricultural production. Unless policy is reoriented towards increasing the comparative advantage of full-time farmers with high entrepreneurial and managerial abilities, it is doubtful that the full structural adjustment in Japanese agriculture now in prospect from changes in technology and labour markets will completely materialize.

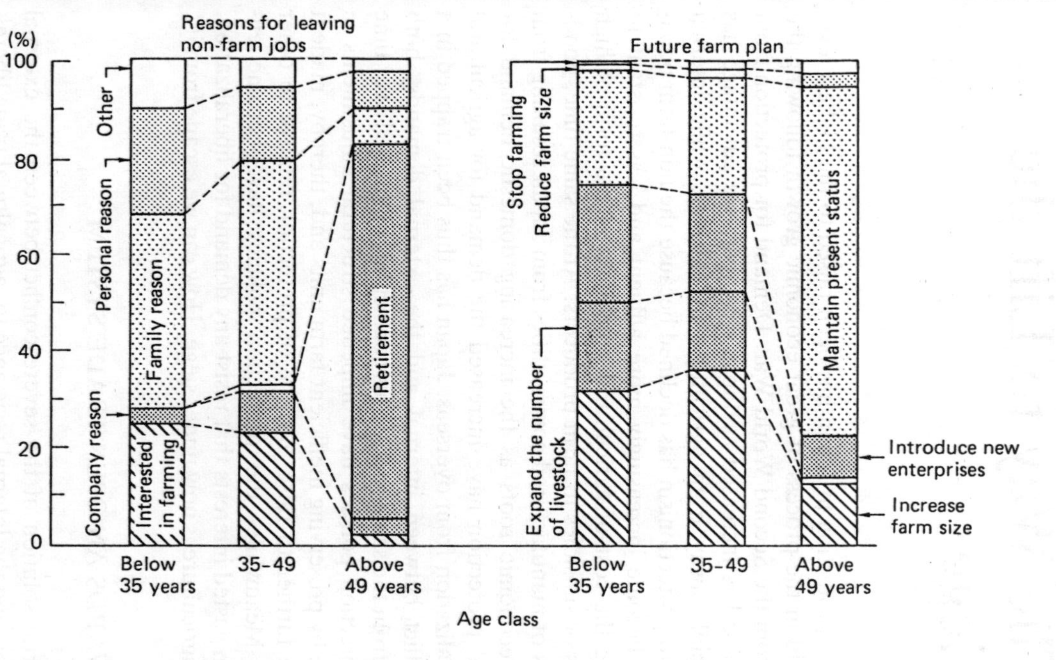

FIGURE 5.6 Reasons why male workers leave non-agricultural jobs for agriculture and their future farming plans (1983)

Source: JMAFF, *Noringyo Census* (Census of Agriculture and Forestry); *Nogyo Chosa* (Agricultural Survey)

6 The Way to Lift the Siege

Agricultural protection in Japan has been raised to the highest level in the world in the process of great economic growth following the recovery from the Second World War. Demand for protection from the farm bloc has increased as Japan's comparative advantage has shifted rapidly away from agriculture to industry, while internal resistance to protectionism has declined because the non-farm population has become increasingly more affluent and less resistant to shouldering the cost of agricultural protection in the form of high food prices or subsidies to farm producers. At the same time, flooding exports of manufactured commodities from Japan resulting from the same economic roots as the increasing domestic demand for agricultural protection have increased the demand for agricultural trade liberalization from overseas. Japan has thus been trapped in a sharp conflict between internal and the external demands, both stemming from the same roots. This conflict has intensified over time as the protection policies have impeded structural adjustments in agriculture by preserving inefficient farm units and, thereby, made it imperative further to raise the rate of protection relative to other countries. Meanwhile, the protection policies have become entrenched in vested interests that resist any demand for liberalization. Japanese agriculture is now under siege. How can we escape from it?

6.1 TOWARDS AN OPEN TRADE SYSTEM

The long-run solution of the severe conflict between the external pressure for, and the internal resistance to, agricultural trade liberalization is to raise the productivity of domestic agriculture to become viable without trade barriers. Japan has already completed the phase of industrial growth based on technology borrowing. The growth rate of industrial productivity is bound to decelerate as its further growth needs greater efforts to push out to new frontiers of technology.

On the other hand, there is still substantial room for technology borrowing in agriculture. In the past, Japan had concentrated its effort in developing agricultural technology in the land-saving and scale-neutral direction. The development of labour-saving technology characterized by increasing returns consistent with the high-wage economy began relatively recently. With appropriate investments in adaptive research and development, Japan should be able to exploit the backlog of labour-saving agricultural technologies accumulated in the US and other high-income countries. A part of the large gap existing between the international frontier and the present level of labour-saving technology in Japan is reflected in the difference in rice production cost among farm size classes corresponding to large-scale mechanization as observed in Chapter 5 (Table 5.1, p. 98).

If this potential is duly exploited, it is not a mere dream to expect that the declining trend in comparative advantage in agriculture can be reversed. There is no reason to assume that Japan will continue to be an exception to a rule that comparative advantage shifts back to agriculture at the highest stage of economic development (Table 1.4, p. 11).

The critical condition for the exploitation of the potential of labour-saving technology is to expand the scale of farm operation. So far, the spread of part-time farming has been inhibiting expansion in the operational scale of full-time farmers. However, this same turn to part-time farming has also helped to equalize agricultural and non-agricultural income levels and has contributed significantly to social stability by preventing rural depopulation and urban overcrowding.

The dual demands for creating an internationally competitive agriculture and preserving social stability can best be met by enabling part-time farmers to stay in the rural villages and to reduce their own operations by consigning production to full-time farmers. Because the farmer's strong attachment to the land and the expectation that such land will become increasingly valuable make it impossible to expand the scale of agricultural operations through transferring land ownership, the only way left to expand operational scale is through land leasing.

Japanese agriculture should thus aim at creating the following features, looking ahead to the next century. First, while guaranteeing highly stable employment opportunities in non-agricultural sectors, it should promote easier transferability of cultivation rights so that some 90 per cent of the 5 million farm families can remain in rural villages, keeping small areas for home gardens, while they work in non-agricultural occupations and consign the cultivation rights of the rest of their land to full-time farmers. If this is done, the approximately 10 per cent of full-time farmers will be able to expand the scale of their operations to an average of about 10 hectares. This is comparable to the West German scale, and – given Japan's high land productivity – would make it possible for Japanese agriculture to compete internationally.

6.1.1 Structural Policy

Activation of the land rental market is essential to the shift to such an agricultural structure. For that purpose, the liberalization of land rental contracts must be promoted further. Major institutional impediments to the transfer of land use rights have already been removed. Yet unnecessary regulations on land tenure still remain – for example, share tenancy is prohibited despite its advantage in reducing risk for active young farmers who try to rent in more farm land as well as facilitating retirement of old farmers by forming partnership with the young. Rent control was removed with the 1970 amendment of the Land Laws, but JMAFF sets a 'standard rent' to which land rental contracts based on the Farm Land Utilization Promotion Law are supposed to conform. Those unnecessary regulations and administrative restrictions should be removed as soon as possible.

Even if an expanded scale of operation is promoted through the activation of the land rental market, this will not in itself mean the achievement of internationally competitive agriculture, given the current production structure. Even if Japan achieves an operational scale on a par with EC levels, it will be difficult to compete internationally in wheat or feed grains with countries in new continents such as the US or Australia. A major reorientation of agricultural production will be required for the shift to an open trade system. The case of the Netherlands (as observed in Table 1.1, p.3) is suggestive. The Netherlands has the lowest self-sufficiency in grains of any European country, even lower than Japan's but at the same time it is a major exporter of horticultural and livestock products because it has succeeded in using its scarce land very efficiently. It has also been successful in maintaining a good pastoral environment in the nation. When cheap grains began to pour into Europe from the New World in the late nineteenth century, France, Germany, and other European powers erected tariff barriers to protect domestic agriculture, whereas the Netherlands and Denmark transformed their agriculture by making positive use of these cheap grains and specializing in production of the commodities that had a comparative advantage.

The agricultural transformation of the Netherlands and Denmark are indicative of a direction Japanese agriculture should take. Of course, we must be fully aware of the differences between the economic environment in which Japanese agriculture operates today and that which existed for the Netherlands and Denmark in the nineteenth century. Japan's industrialization has been very rapid in comparison with Europe's in that century, and thus our transition period must necessarily be shorter. At the same time, it will not be easy for Japanese farmers accustomed to agriculture concentrating on traditional rice and cereal crops to switch to livestock-centred farming and to bring their productivity up to international standards. Greater governmental investment in agricultural research, education and training of farmers and land infrastructure will therefore be needed if the desired agricultural adjustments are to be achieved.

6.1.2 Rice Policy

A major obstacle to such a restructuring of Japanese agriculture is the high level of rice price supports. The price of rice paid to Japanese farmers was virtually at the international level until the early 1950s. It was raised sharply during the era of rapid economic growth. During the period when the producers' price for rice soared, investments in technological development, extension services, and land improvement were concentrated on rice production to make rice Japan's most stable and most standardized crop. The result is that part-time farmers have come to rely upon rice (Table 4.8, p. 89). Meanwhile, there has been little expansion in the scale of operation of full-time farmers. The shift of agricultural resources from rice to other crops with high income elasticities has been blocked. Deficits in the rice-dominated Food Control Account, together with the costs involved in the control of areas planted in rice, have come to account for a major share of national budget for agriculture during the past two decades (Table 3.6, p. 63). The fact that the Food Control Account has eaten up a large portion of the government's expenditure for agriculture has made it difficult to increase investments for agricultural restructuring.

The price of rice must therefore be held down before structural adjustments can be made that would lead to an open trade system. Although lower rice prices may not be welcomed by farmers, they will ultimately benefit even those full-time farmers who specialize in rice cultivation by promoting the transfer of cultivated land and, thereby, enabling them to expand their scale of operation. Holding down the price of rice is also needed to encourage the shift away from rice – where demand has been declining from 2–3 per cent per annum – to other crops where demand is increasing.

This shift away from rice cultivation must be promoted with a continuing concern for a viable agricultural structure under an open trade system. The shift to barley or soybeans is now encouraged by the government. However, such a shift represents a most inefficient resource reallocation in terms of comparative advantage. Rather, the main shift must be to forage crops as the basis of livestock production, which is expected to continue a rapid demand growth. As can be seen in Table 6.1, the rates of increase in total food consumption (as well as in individual food items) declined rather sharply from before to after the First Oil Crisis as the growth of the Japanese economy was decelerated. But the consumption of meat and milk (and dairy products, especially cheese) has been rising relatively rapidly. Table 6.2 shows that Japanese intake of animal protein is still low by international standards. Because it is difficult to hope for any major increase in fish production under the present economic and technological conditions, demand for meats, milk, and dairy products will most probably continue to grow rapidly.

Although it may be difficult to shift profitably from rice to forage crops, given the current state of technology in Japan, it should be possible to develop highly efficient mixed farming if the operational scale can be expanded and appropriate research and extension activities promoted.

6.1.3 Price and Trade Policies

Concerted efforts for such structural adjustments seem to be the best way to adapt Japanese agriculture to an open trade system in the long run. At the same time, the international pressures on Japan make it impossible to delay the liberalization of agricultural imports until structural adjustments are completed. It is thus necessary to formulate price and trade policies to cope with the urgent trade friction.

A problem with Japanese agricultural protection is that it is creating an international image of unfairness worse than reality because it

TABLE 6.1 Changes in food consumption in Japan

	Quantity consumed						Rate of change (%/Year)	
	1960	65	1970	75	1980	85	1960/ 1970	1975/ 1985
Consumption *per capita* per day:								
Calorie	2290	2459	2529	2516	2554	2581	1.0	0.3
Protein (g)	70	75	78	80	83	84	1.1	0.5
Animal protein (g)	21	26	31	35	39	41	3.9	1.6
Consumption *per capita* per day:								
Grains	150	145	128	122	113	109	–1.6	–1.1
Rice	115	112	95	88	79	75	–1.9	–1.6
Vegetables	100	108	114	110	110	108	1.3	–0.2
Fruits	22	29	38	43	39	37	5.5	–0.3
Meat	5	9	13	18	23	25	9.6	3.3
Beef	1.1	1.5	2.1	2.5	3.5	4.4	6.5	5.7
Egg	6	11	15	14	14	15	9.2	0.1
Milk and dairy products	22	38	50	53	62	67	8.2	2.3
Fresh milk	13	20	26	29	34	35	6.9	1.9

Source: JMAFF, *Shokuryo Jikyu Hyo* (Food Balance Sheets).

TABLE 6.2 Comparison of protein intake level between Japan and other industrial countries (g/day/person)

Country	Animal protein					
	Meat		Eggs and dairy products	Fish	Total	Total protein
	Beef	Total				
1982:						
USA	19.8	40.0	30.4	2.3	72.7	102.4
Canada	17.2	33.4	31.2	2.4	67.0	98.4
UK	8.4	25.1	27.6	4.0	56.6	88.2
France	12.8	38.7	36.1	4.5	79.3	114.2
Germany (FR)	8.9	31.9	33.0	2.7	67.5	98.8
Italy	10.5	27.7	26.9	2.9	57.5	103.7
1985:						
Japan	2.2	12.4	10.3	18.5	41.3	84.1

Source: JMAFF, *Shokuryo Jikyu Hyo* (Food Balance Sheets); OECD, *Food Consumption Statistics*.

relies, to a large extent, on import quotas (IQ) in violation of the GATT Code and stringent state-trading monopoly. The quota protection has raised the domestic prices of several items such as beef and oranges so far out of line with international prices that they have become symbols of Japanese protectionism. This is an undesirable situation, detrimental to both our international relations and domestic consumer interests.

The immediate step Japan should take in coping with the trade friction is to abolish all the IQ restrictions. In designing the removal of the remaining 22 items from the IQ list (Table 3.1, p. 53), it is useful to classify them into three categories. The first group includes minor agricultural products of mere local significance, such as beans and peanuts. The decline in these crops corresponding to trade liberalization may have serious impacts on specific local economies, but the same applies to mines and factories in the declining industries. The standard approach to the liberalization of such commodities – either agricultural or manufactured – is to assist industrial adjustments, such as giving subsidies to the producers for the shift of resources to other economic activities while inducing the location of new enterprises in substitution of the declining ones by giving tax preference and subsidies.

The second group includes the products of sufficient international competitiveness. Japanese *mikan* (tangerine) production is representative of this group. For the 1982–5 average, the wholesale price of *mikan* was about 30 per cent lower than the cif price of imported oranges, and only half that if the seasonal duty (applicable in December to May) of 40 per cent *ad valorem* is added. Given that much of a price difference, it is inconceivable that the import of oranges will increase so much as to imperil domestic *mikan* production. The removal of the IQ restriction on commodities like oranges may begin by adding a certain levy on the top of the standard tariff, and cut it down gradually over several years while using the levy revenue for the compensation of marginal producers.

In the case of oranges, rather than holding out for the negative policy of simply resisting import liberalization, it would be better to initiate a positive policy of seeking the *quid pro quo* of expanded US imports of Japanese *mikan* (now restricted for plant quarantine reasons) and trying to expand both domestic and international demand by mixing imported orange or grapefruit juice with *mikan* juice for a better-quality, higher-value product. There are also many possibilities besides juice where the liberalization of agricultural imports can be made the source of an increase in exports of processed goods.

The third group may be represented by beef, which is now weak in international competition but is expected to become a major industry in Japanese agriculture in the future, especially in the remote hills and mountains. For a product like beef – where the domestic industry is still in an infant stage and where demand is elastic enough that liberalization can be expected to generate substantial increases in demand – there is every possibility that a combination of import liberalization and deficiency payments could meet the dual aims of promoting international cooperation and sustaining the maintenance and development of the domestic industry (Hayami, 1979; Longworth, 1983, pp. 294–8; Sanderson, 1983).

While the first priority should be given to the abolition of IQ protection, a long-run design of liberalizing the trade of commodities now under the control of state trading must be prepared. Even in the case of rice – where trade liberalization is likely to be very remote – steps must be taken to reduce the gap between domestic and international prices under the grand design of structural adjustments to achieve viable agriculture under the open trade system in the next century.

6.1.4 Towards Market-oriented Policies

The efforts of structural adjustments are bound to fail unless the institutional framework of domestic agriculture is reformed so as to allow the working of a competitive market mechanism. The essential step towards viable agriculture is the concentration of productive resources in the hands of farmers who have high entrepreneurial ability. A major impediment to this goal is government regulations and controls on agricultural product and input markets.

Again, rice is the typical example. An important reason why a major share of rice output is produced by part-time II farmers is the government controls on its price and marketing channels. Farmers who produce commodities under free market conditions (such as vegetables) must always adjust production and marketing in response to price fluctuations, the success of marketing is a critical determinant of the profitability of their enterprises. In contrast, there is no room for rice farmers to take advantage of their entrepreneurship in this regard, because the price is fixed and the produce is collected almost exclusively by agricultural cooperatives (*NOKYO*). Part-time farmers who work in factories and offices can wait to harvest and deliver their rice crop until the weekend with no loss in profit. In that sense, the rice control programmes have the effect of lowering the relative advantage of farmers with high entrepreneurial ability working full-time in agriculture.

The present system of acreage control needed to maintain the rice price far above the domestic equilibrium under autarky also has the effect of blocking the concentration of production in the hands of efficient producers. In the present system, subsidies are paid to rice producers who divert paddy fields for the production of other crops, but are not sufficient to induce their voluntary diversion because the high price support makes it too lucrative to grow rice. Therefore the acreage control is imposed by government on all producers more or less at a uniform percentage of their paddy fields. In order to enable the exploitation of scale economies by efficient producers it is necessary to change the present system to a system of voluntary participation as is practised in the US, in which the government rice purchase at a guaranteed price is limited to those who participate in the acreage control programme, and the rest are free to produce any amount and sell it at any place while taking the risk of price fluctuations.

The merit of this system is not limited to the concentration of production among efficient producers. Producers who do not prefer to participate in the acreage control programme need not sell their produce to government-designated collectors (*NOKYO*). Consequently, competition will emerge between *NOKYO* and private traders, and those who offer the best service to producers will increase their share of the market.

Improvements in marketing services through competition must be promoted in the supply of farm inputs, too. At present, *NOKYO* has a dominant share in the supply of fertilizer and other inputs that were rationed through *NOKYO* during the Second World War. Moreover, because much of the government subsidies and low-interest loans related to the purchase of farm inputs is channelled through *NOKYO*, the entry of private traders and manufacturers into the marketing of agricultural inputs tends to be blocked.

A condition of the development of viable agriculture will be the provision of marketing services consistent with the demands of entrepreneurial farmers through healthy competition between *NOKYO* and private dealers. The competition will not be limited to price competition, but will be expanded to competition with respect to the provision of information on market, technology and management know-how. In order to promote the competition, the Anti-Trust Laws that exempt *NOKYO* from their application must be reconsidered. Also, the regional franchise of *NOKYO* by village should be reformed so that farmers become free to choose among several cooperatives that compete in a same district.

The critical condition for the development of viable agriculture is the growth in entrepreneurial ability of farmers. It can hardly be expected that such human capital will develop unless the institutional framework is such that entrepreneurial ability is well rewarded. In that sense, the restoration of a competitive market mechanism is the key to the development of viable agriculture.

6.2 THE MYTHOLOGY OF PROTECTIONISM

In order to reorient Japanese agriculture towards an open system, it is necessary to demolish in public eyes several myths used for the justification of protection policies.

6.2.1 The Fallacy of Food Security

Until now, concern for food security has been used as a major justification to ward off liberalization and to promote food self-sufficiency. Yet the logical basis of such an argument is fragile indeed.

In discussing Japan's food security, it is first necessary to define the possible crisis which may occur. These may be broadly classified into: (a) diminished supplies and higher prices for foods as a result of poor harvests worldwide, such as happened in 1973–4, (b) a halt to imports because of war or some other catastrophe, and (c) a Malthusian crisis which might emerge if population grew relative to food supplies.

For the first case, it would not be very effective to increase the food self-sufficiency rate ignoring considerations of comparative advantage. The US embargo on soybean exports is a frequently cited example of such a crisis. Yet raising Japan's self-sufficiency in soybeans from about 4 per cent at present to a future 10 per cent would do very little to alleviate the panic caused by such a crisis. Moreover, any attempt to raise Japan's self-sufficiency to a meaningful extent would necessarily entail socially unacceptable costs. Effective policies to cope with such a crisis would be to diversify import sources and to cooperate with international stockpiling programmes.

Nor would the maintenance of domestic self-sufficiency be at all effective in a crisis of the second type, a war-induced or other halt to imports. Because Japanese agriculture is heavily mechanized, and because any situation that would cut off food imports would also cut off oil supplies, peacetime self-sufficiency levels would be irrelevant in coping with such a crisis. The first means for preparing for such a crisis would be to enlarge domestic stockpiles. Expanded livestock farming would also be a very effective contingency step, in that livestock can be slaughtered for meat in times of shortage and the operational stock of feed grains can itself be used for food. Yet it is essential that the policies designed to survive such a crisis be implemented before the crisis hits us. The seeds, materials, and manpower mobilization plans must be ready so that food rationing can be instituted, and pastures, golf courses and other areas be organized for cultivation as soon as the crisis strikes.

To deal with the third possible crisis (a Malthusian food shortage) does require that Japan's agricultural production capability be maintained and enhanced. However, the need is for potential productive capacity, and not for continued production of products with which Japan is at a comparative disadvantage. So long as irrigation facilities and other land improvements are kept up, the land can be used for forage crops now and can be easily converted to grain production later should the situation demand. Agricultural research and development must also be promoted from the long-term point of view. Although domestic production of barley, soybeans, and other crops may be inefficient at present, this situation could change if the global supply-demand balance deteriorated sharply in the future. Our long-term security needs thus demand that research be steadily pursued in preparation for such a contingency. Above all, the best policy for coping with such a Malthusian crisis is to cooperate with agricultural development in developing countries, thereby forestalling the crisis. Agricultural research and development are the most effective means for such cooperation.

Ensuring food security is one of the government's most important responsibilities, and food policies must be forcefully promoted to this end. However, to tie food security to a short-term improvement in domestic self-sufficiency not only impedes preparations for an open trade system but carries the very considerable risk of diverting attention from the programmes that need to be undertaken for true security.

It is most important to recognize that the national security of a country like Japan that renounces armaments by its constitution can be maintained only by international cooperation. Food is no exception to this rule. If food self-sufficiency is promoted at the expense of international friction, it has only a negative effect. How the promotion of food self-sufficiency with disregard to international harmony is ineffective in crisis was evident from the experience during the Second World War. Before the war, the Japanese Empire including Korea and Taiwan boasted the self-sufficiency of food staples. How that was effective in food security is fresh in the memory of all the Japanese who experienced hunger during the war. The following criticism of the food security argument before the war is directly applicable to that of today:

> But, it was not duly explained why food alone is considered so important in crisis. The crisis in the modern world can hardly be coped with by such low economic power as to secure food alone . . . It was surprising that a disproportionately large weight was attached to food in the design of total war. When a country faces a crisis in the food supply, all other aspects of the economy are in crisis. *A country is in danger not because food becomes short of supply. Rather, when the country is in danger, food becomes short of supply* . . . It was a farce to observe that those who voiced loudly for food self-sufficiency were wearing clothes made in wool and cotton imported from abroad (Tobata, 1956, p. 597, author's italics).

6.2.2 Are Farmers the Victims of Economic Growth?

Another justification of protection policies argues that farmers are the victims of high economic growth and that, therefore, they must be protected. Besides the question of whether trade protection is an efficient method of compensation, it is clear that, on the average, farmers are not victims but major beneficiaries of high economic growth. As was observed in Table 4.9 (p. 92) the average *per capita* farmhousehold income in 1960 was about 30 per cent lower than the *per capita* urban worker household income, but the disparity had been narrowed during the process of high economic growth until the former exceeded the latter in 1975. The main force underlying this rapid increase in farm household income was the increase in off-farm income that now occupies more than 80 per cent of total farm household income, though agricultural protection made a marginal contribution, too.

The fact that farmers are the major beneficiaries of economic prosperity means that they will become the victim of economic recession. In fact, relative to urban households, non-farm employment of farm family members is less stable and more subject to layoffs. Their income is therefore more responsive to economic fluctuations than the income of urban dwellers. Once this fact is recognized, it should be the vital interest of farm population to maintain the free trade regime on which Japan's economic prosperity has been based.

Of course, the benefit of free trade spills over to other groups besides the farm population. A part of adjustment cost of agricultural trade liberalization should therefore be shouldered by the non-farm population. If that compensation is wisely used for the provision of public goods (such as agricultural research and extension to improve farm production efficiency, education and training to assist farmers and their family members in finding more lucrative non-farm employment opportunities, and insurance and retirement pension for elderly farmers), it will have a much higher rate of return to them than the protection policies.

6.2.3 Conservation of the Natural Environment

Agricultural protection has been justified also on the ground that agricultural activities contribute to the conservation of the natural environment for the benefit of the whole nation. For example, it has frequently stated that rice production in remote hills and mountains should be maintained even if it does not pay economically, because terraced paddy fields in the hillsides have a function of water reservoirs for the prevention of floods in the low land.

While this argument is valid, it does not justify the use of trade restriction or price support policies. Activities characterized by external effects – such as rice farming in the hillsides – alone may be subsidized. If trade restriction or price support is used for the support of marginal farmers in the hillsides, it will create a large rent in superior low land. That is not only unfair, but also detrimental for structural adjustments, because the high rent discourages the scale expansion of efficient producers.

It has also been argued that small subsistence-oriented peasants should be preserved because of their ability to recycle resources with little emission to create environmental pollution. However, few of such subsistence-oriented peasants exist in Japan any longer. Small producers are mainly part-time II farmers. Because the opportunity cost of their labour is high, they rely heavily on modern chemical inputs for the labour-saving purpose. In fact, large commercial farmers are making much more effort than small part-time farmers to conserve soil fertility with the application of organic materials.

6.3 BARRIERS TO POLICY REORIENTATION

Hidden under the mythology, agricultural protection policies are entrenched deeply in vested interests – above all, the powerful *NOKYO* organization with its alliance with JMAFF and LDP stands ready to fight against any move towards trade liberalization. Even if the total income of farm households may increase by trade liberalization through an increase in their off-farm income, that is of no benefit to *NOKYO*.

Explicitly or implicitly, *NOKYO* resist even structural adjustment policies geared to raising the productivity of Japanese agriculture to an international level. In general, small part-time II farmers are obedient customers of *NOKYO*, because the opportunity cost of their labour is too high to spend time in the search of better marketing opportunities for their products and inputs other than *NOKYO*. After all, their income from farming is less than 10 per cent of their total income. In contrast, large entrepreneurial farmers try hard to search and bargain for the best deal. The successful transformation of the agricultural structure from one dominated by a large number of small part-time farmers to one dominated by a small number of large entrepreneurial farmers would be a nightmare to *NOKYO*.

In particular, the *NOKYO* organization resists strongly any move to restore a competitive market mechanism. Their major effort has been concentrated on the preservation of the food control system by which the entry of private traders to rice marketing is strongly restricted. Also, *NOKYO* is an ardent supporter of the domestic fertilizer cartel based on the Fertilizer Price Stabilization Act, which has imposed on farmers a price of fertilizers much higher than the international price.

Probably, the greatest myth of Japanese agriculture today is the slogan that *NOKYO* is 'of the farmers, by the farmers and for the farmers'. How can *NOKYO* continue to organize farmers for the preservation of their vested interests, even at the expense of farmers? A major reason is that the political pressure of *NOKYO* is exerted mainly to induce the policies that bring about overt short-run benefits for farmers, while their negative effect is not so visible. For example, the damage on *mikan* producers from the removal of IQ restriction on orange imports is obvious even though it will be very modest, the benefit from the trade liberalization through the increase in off-farm income of farm households is too indirect and thinly diffused to be visible. The high price of fertilizers that *NOKYO* sell is also invisible because no private trader and manufacturer dares to sell fertilizers cheaper than the cartel price for fear of retaliation by *NOKYO*, which has an overwhelming share in fertilizer marketing. In contrast, government subsidies that are often channelled through *NOKYO* are visible. Moreover, *NOKYO* usually has some discretion in allocating the subsidies in favour of cooperative members.

In addition, *NOKYO* leaders are usually influential persons in local communities, and it is not easy for a villager to take an independent action in defiance of the *NOKYO* interests, especially in the 'tightly-structured' village communities of Japan characterized by comformity to social norms (Embree, 1950). This tendency is augmented by the strong egalitarianism of village people, who feel uncomfortable where a specific villager becomes competitive in the market and expands his farm. *NOKYO* leaders take advantage of this psychology to prevent the introduction of the market mechanism. It is an irony that the tightly-structured village communities consisting of relatively homogeneous peasants, which were a highly effective instrument for developing and diffusing land-saving and yield-increasing technologies without impairing social equity in the low-income stage of development, have now turned into a bastion of inefficiency and vested interests.

It is fair to say not all the *NOKYO* leaders make negative contributions. Many of them are conscientious individuals, caring for the wellbeing of their fellow villagers. Some of them are making truly innovative contributions to community development programmes. Yet there is no denying that the *NOKYO* organization as a whole is today a major seeker of 'institutional rent' arising from government regulation and control (Tullock, 1969; Stigler, 1971; Krueger, 1974; Bhagwati, 1982; Tollison, 1982).

It should also be pointed out that Japanese farmers are not unique in being misled by the rent seekers into moving against their own interests. In nineteenth-century Germany under Bismarck, powerful Junkers in Prussia lobbied fiercely for tariff protection on grains produced on their estates. Strangely, medium and small peasants in the western part of Germany also sided with this campaign under the slogan of 'agricultural protection', despite the fact that the imports of cheap grains were beneficial to them because, unlike the Junkers, their operation depended more heavily on livestock than grain production (Gerschenkron, 1943, pp. 28–32).

It is a common observation that the political power of the farm bloc rises in parallel with the relative shrinkage of the agricultural sector. An international comparative study shows that it becomes the strongest when the share of agriculture in total labour force ranges from 5 to 10 per cent (Anderson and Hayami *et al.*, 1986, Chapter 4). Some industrial countries such as the US, the UK and Germany have passed through this range, while Japan is approaching it. It is unlikely that the farm bloc will be weakened significantly for some time to come. The political clout of the 'iron triangle' consisting of *NOKYO*, JMAFF and ruling LDP may be able to block out increasing foreign demands for agricultural trade liberalization, even at the expense of national interests arising from possible retaliations from the trade partners.

Under the guard of the 'iron triangle', can Japanese agriculture survive? Because the 'voice' from abroad is by nature a political action, as Hirschman (1970) defines it, it may be counteracted by the political action of the farm bloc. However, no political action can stop the 'exit' of customers in the market. The large gap between domestic and international prices due to trade barriers represents a strong pressure to foreign agricultural products to leak into the Japanese market in some way or other. A major gap for such a leakage is the liberalized trade of processed farm commodities. While the imports of rice and wheat are severely limited, spaghetti from Italy, biscuits from England, *arare* (rice cookies) and *mochi* (rice cake) from Thailand are flooding into the Japanese market. Japanese consumers are thus indirectly exiting from the domestic agricultural products. This trend will be accelerated as the food-processing industries increasingly shift their production base overseas.

If the 'iron triangle' continues to be successful in warding off foreign pressure, Japanese agriculture will shrink steadily through the vicious circle of high domestic prices and the exit of consumers. In the end, the economic base of *NOKYO* will also be lost.

This exit factor operates not only in the product market but also in the labour market. If *NOKYO* continues to resist the restoration of a competitive market mechanism, farm boys with high entrepreneurial ability will leave agriculture for other industries where their ability will be more highly rewarded. Inevitably human capital in agriculture will continue to deteriorate.

Since the exit is a market response, the only way to counteract it is to improve services to customers in the market. It is therefore vital for the survival of Japanese agriculture to reduce the prices of domestic agricultural products and to prepare the institutional environment for raising the returns to farmers' entrepreneurship. Policies are needed to maximize the speed of structural adjustment side by side with steady liberalization in the domestic market and in international trade.

Indeed, Japanese agriculture is now at the crossroads. Can Japanese agriculture avoid taking a route to progressive decay under the siege of international criticism? Or can it restructure itself into a viable economic sector in international harmony? Very soon we will reach a point of no return.

Appendix A: A Model of the Political Market for Agricultural Protection

In terms of the economic theory of politics (Downs, 1957; Buchanan and Tullock, 1962; Breton, 1974), demand for and supply of agricultural protection may be considered the schedules of marginal revenue and marginal cost to political entrepreneurs or leaders expected from instituting policies to raise the level of protection. Changes in the equilibrium level of protection corresponding to shifts in the marginal revenue and the marginal cost curves are illustrated in Figure A.1. The horizontal axis represents the protection rate as measured by the rate of increase in agricultural income resulting from protection policies, including border protection, direct price support and subsidies. O represents the point at which the protection rate is zero.

The vertical axis represents the marginal gain and the marginal loss that politicians expect to incur if they would raise the protection rate by 1 additional unit. If we follow Downs in assuming parliamentary democracy, the gain may be measured by the expected increase in the number of votes in support of the politicians from the group demanding protection policies, while the loss may be measured by the expected decrease in votes from the opposition group. More generally, the marginal revenue and the marginal cost curves in Figure A.1 may be considered politicians' marginal evaluations of expected increase and decrease in political support related to their probability of staying in power, irrespective of whether the political framework is democracy or not.

It would be reasonable to assume that the marginal revenue curve is downward-sloping because the intensity of a political campaign for protection is likely to diminish at a higher level of protection, while the resistance is likely to multiply, resulting in the rising marginal cost curve. The subjective equilibrium of maximum expected profit (revenue minus cost) for politicians will be established at the intersection between the marginal revenue and the marginal cost curves.

In the early stage of economic development the majority of people engage in agriculture. However, most of the agricultural population are uneducated and live sparsely over a wide area with poor communications and transportation

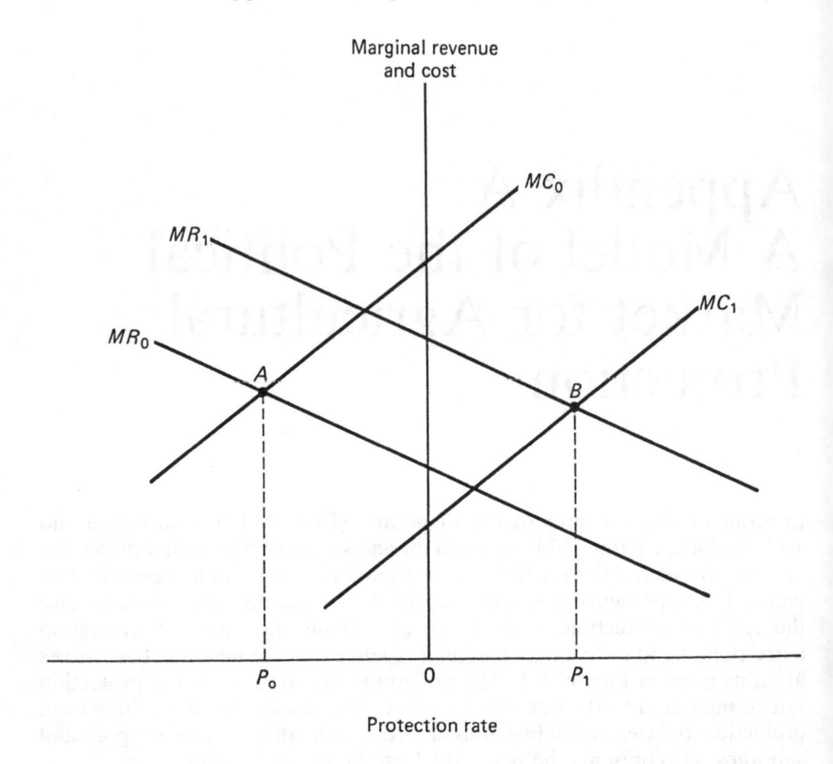

FIGURE A.1 Transition from the agricultural exploitation to the agricultural protection policy

infrastructure. They are largely ignorant about the adverse effects of covert exploitation policies such as export taxes and overvalued exchange rates. Even if they do know, it would be too costly to organize the counter-acting group action (Olson, 1985). Moreover, because social instability and disruption in remote villages are not an immediate menace to the present regime, politicians tend to put a low weight in their profit calculation on the demand of rural dwellers. For those reasons, the marginal revenue of agricultural protection in the early development stage is likely to be low, which may be presented by MR_0 in Figure A.1.

At an early stage, the population engaging in modern industries is small but concentrated in the metropolis. Because their incomes, are still low and their Engel coefficient high, high food prices tend to trigger urban disruption that would immediately imperil the government. Business elites who are small in number but well organized strongly resist the increase in food prices, for fear of the increased cost of living and the resultant high wages of their workers. Intellectuals tend to accept the policies to promote industrialization at the expense of agriculture, because they are keen for their nation to catch up the advanced economies through industrial development. In such a society, the marginal costs of protection are likely to be high, as represented by MC_0.

Thus, in an economy in which the food problem dominates, the agricultural protection rate is typically negative as represented by OP_0 corresponding to *A* (at which the expected profit of politicians is maximized).

The equilibrium protection rate will change as the economy develops. While the farm population declines, their educational level rises and the communication and transportation infrastructure in rural areas improves. Rural people become more sensitive about their position relative to urban dwellers in income and the level of living, while the cost of organizing themselves declines. Under such conditions, if comparative advantage in agriculture declines and the income position of farmers deteriorates, they begin to resort to political action to create urban–rural equity. Correspondingly, the marginal revenue curve of protection for politicians will shift upwards from MR_0 to MR_1.

At the same time, as the food problem loses ground due to increases in *per capita* income associated with increases in the Engel coefficient, urban dwellers become less resistant to high food prices. People become more tolerant of the high cost of agricultural protection as their nostalgia for a pastoral life increases and their interest in environmental conservation rises. Correspondingly, the marginal cost schedule declines, as represented by a shift from MC_0 to MC_1.

Thus, as the food problem is replaced by the agricultural adjustment problem in the course of economic development, the optimum level of agricultural protection in politicians' calculation will change from negative (OP_0) to positive (OP_1). This process involves both coalition and competition among interest groups such as landlords and industrial capitalists (Bates and Rogerson, 1980; Becker, 1983), while changes in the sizes of those groups alter their relative efficiencies in organizing their political lobbying (Olson, 1965). An econometric investigation confirms that this political market model is consistent with the experience of agricultural protection growth in Japan, in which no unique bias specific to Japan – such as agricultural fundamentalism and special preference for food security – for preserving domestic agriculture is found (Honma and Hayami, 1986; Anderson and Hayami *et al.*, 1986, Chapter 4).

Appendix B: A Model of the Rice Policy in Japan

The interactions between market forces and government rice policies as explained in chapter 3 may be represented by the Marshallian diagram in Figure B.1. The vertical and the horizontal axis measures, respectively, the price and the quantity of rice traded in market. The domestic demand schedule (DD) is drawn as a steeply sloped curve in order to represent the low demand elasticity of rice. Domestic supply (SS_0) is drawn as a less steep curve, because it is intended to represent a long-run supply response allowing adjustments in fixed capital investments including conversion of dry upland to irrigated paddy fields.

OP_w represents the international price. If no border protection exists and the assumption of 'small country' applies, the production equilibrium of the domestic market will be established at M and the consumption equilibrium at N with the quantity of import at the level of MN. The domestic price can be raised by means of border protection up to OP_e, at which domestic demand and supply are equated. It appears that the rice market in Japan was at such autarky equilibrium in the early 1960s, although the price was determined not by the market but by the government.

However, as the price has been raised since then beyond the autarky equilibrium, excess supply has inevitably emerged. As the consumer price is raised to OP_c, the consumption equilibrium is established at C. Domestic supply is considered highly inelastic in the short run. Therefore, even if the producer price is raised to OP_p higher than the producer price, the quantity of rice supply would remain not so different from OQ_e in the short run. The government has to shoulder for each unit of rice output the difference between the consumer price (OP_c) and the producer price (OP_p) plus the government marketing cost (P_gP_p). This current account deficit from the rice control programme can be reduced if the consumer price is raised with a modest time lag, as was actually done. Therefore, the government deficit does not reach a serious level so long as the excess supply remains relatively small.

As time passes, however, the production equilibrium will shift to the long-run equilibrium point (E) with the excess supply increased to the level of Q_cQ_p. The excess supply is added annually to a surplus government inventory and will have to be disposed of eventually at the salvage price

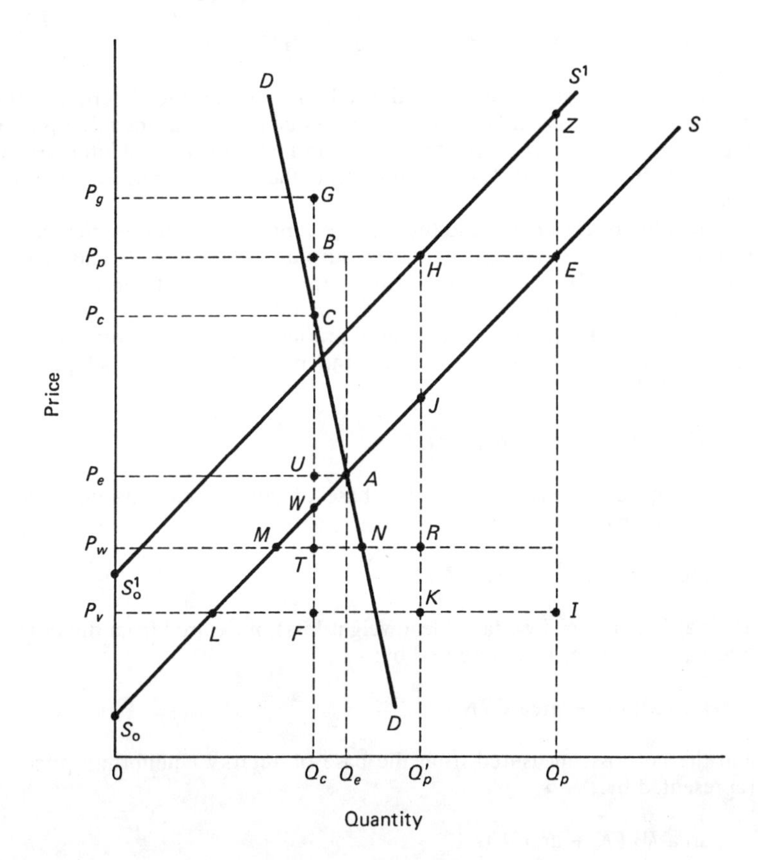

FIGURE B.1 Effects of the rice policy in Japan

(OP_v) which is usually lower than the international price. The capital loss arising from the disposal of old rice is represented by the area *EBFI*.

The acreage control is made necessary by the need to reduce the large cost of old rice disposal. Subsidy payments for the acreage control have the effect of shifting the supply curve upward. For the sake of expository simplicity, assume that a fixed amount of subsidy is paid for each unit decrease in rice output. Then, the marginal cost curve for producers – and, hence, the market supply curve – shift in parallel from SS_0 to $S'S'_0$, as shown in Figure B.1.

Correspondingly, the production equilibrium point moves to *H*, and the excess supply and its disposal cost are reduced to Q_cQ_p' and the area *HBFK*, respectively. However, this policy requires the subsidy payment to rice producers by area *EZHJ*.

The total budget cost for the rice price support programme, including the acreage control, is represented by:

area GP_gP_cC + area $HBFK$ + area $EZHJ$.

The total social cost can be obtained by adding the consumers' burden to this budget cost. The consumer burden, or the reduction in consumers' surplus, is represented by area CP_cP_wN, if it is measured from the base of international price, and by area CP_cP_eA, if measured from the base of autarky equilibrium price.

Under the price support and the acreage control programmes, the income of producers is the sum of producers' surplus (area HP_pS_0J) and subsidy receipt (area $EZHJ$). Their income in the free market condition will be only the area MP_wS_0 under complete trade liberalization and the area AP_eS_0 under autarky. Therefore, the benefit of producers from the government rice control programmes, measured from the base of international price, is represented by:

area HP_pP_wMJ + area $EZHJ$

and the benefit measured from the base of autarky equilibrium price is represented by:

area HP_pP_eAJ + area $EZHJ$

The net loss in social welfare (deadweight loss), measured from the base of international price, is represented by:

area $JWFK$ + area CTN

and the net loss, measured from the base of autarky equilibrium price, is represented by:

area $JWFK$ + area CWA

The above analysis assumes away several complications, such as producers' home consumption and the voluntary rice channel. However, the basic conclusions remain unchanged even if those complications are included (Otsuka and Hayami, 1985).

Bibliography

Anderson, Kym and Hayami, Yujiro *et al.* (1986) *The Political Economy of Agricultural Protection: East Asia in International Perspective*, Sydney, London and Boston, Allen & Unwin.

Bale, Malcolm D. and Lutz, Ernest (1981) 'Price Distortions in Agriculture and their Effects: An International Comparison', *American Journal of Agricultural Economics*, 63, 1, February, pp. 8–22.

Bates, Roberts H. (1981) *Markets and States in Tropical Africa: The Political Basis of Agricultural Policies*, Berkeley, Los Angeles and London, University of California Press.

Bates, Robert H. and Rogerson, William P. (1980) 'Agriculture in Development A Coalition Analysis', *Public Choice*, 35, 5, pp. 513–27.

Becker, Gary S. (1974) 'A Theory of Social Interactions', *Journal of Political Economy*, 82, 6, November–December, pp. 1063–93.

Becker, Gary S. (1983) 'A Theory of Competition among Pressure Groups for Political Influence', *Quarterly Journal of Economics*, 98, 3, August, pp. 317–400.

Bhagwati, Jagdish N. (1982) 'Directly Unproductive Profit-seeking (DUP) Activities', *Journal of Political Economy*, 90, 5, October, pp. 988–1002.

Breton, Albert (1974) *The Economic Theory of Representative Government*, Chicago, Aldine.

Buchanan, James M. and Tullock, Gordon (1962) *The Calculus of Consent*, Ann Arbor, University of Michigan Press.

Dore, Ronald P. (1958) *Land Reform in Japan*, London, Oxford University Press.

Downs, Anthony (1957) *An Economic Theory of Democracy*, New York, Harper and Row.

Embree, John F. (1950) 'Thailand – A Loosely Structured Social System', *American Anthropologist*, 52, 2, April–June, pp. 181–93.

Francks, Penelope (1984) *Technology and Agricultural Development in Pre-War Japan*, New Haven and London, Yale University Press.

Gerschenkron, Alexander (1943) *Bread and Democracy in Germany*, Berkeley and Los Angeles, University of California Press.

Gerschenkron, Alexander (1962) *Economic Backwardness in Historical Perspective*, Cambridge, Mass., The Balknap Press of the Harvard University Press.

Hayami, Yujiro (1972) 'Rice Policy in Japan's Economic Development', *American Journal of Agricultural Economics*, 54, 1, February, pp. 19–31.

Hayami, Yujiro (1979) 'Trade Benefits to All: A Design for the Beef Import

Liberalization in Japan', *American Journal of Agricultural Economics*, 62, 2, May, pp. 342–7.

Hayami, Yujiro (1982) 'Adjustment Policy for Japanese Agriculture in a Changing World', in Emery N. Castle and Kenzo Hemmi (eds), *US–Japanese Agricultural Trade Relations*, Baltimore and London, Johns Hopkins University Press, pp. 368–92.

Hayami, Yujiro (1986a) 'Agricultural Protectionism in the Industrialized World: The Case of Japan', paper presented at the Conference on Agricultural Protectionism in the Industrialized World, held in the East–West Center, Honolulu, 17–21 February, organized by the National Center for Food in Agricultural Policy in the Resources for the Future Inc., Washington, DC.

Hayami, Yujiro (1986b) *Nogyo Keizairon* (Agricultural Economics), Tokyo, Iwanami Shoten.

Hayami, Yujiro with Akino, Masakatsu, Shintani, Masahiko and Yamada, Sabro (1975) *A Century of Agricultural Growth in Japan*, Tokyo, University of Tokyo Press; Minneapolis, University of Minnesota Press.

Hayami, Yujiro and Kikuchi, Masao (1981) *Asian Village Economy at the Crossroads*, Tokyo, University of Tokyo Press and Baltimore, Johns Hopkins University Press (1982).

Hayami, Yujiro and Ruttan, Vernon W. (1970) 'Korean Rice, Taiwan Rice, and Japanese Agricultural Stagnation: An Economic Consequence of Colonialism', *Quarterly Journal of Economics*, 84, 4, November, pp. 562–89.

Hayami, Yujiro and Ruttan, Vernon W. (1985) *Agricultural Development: An International Perspective*, revised edn, Baltimore and London, Johns Hopkins University Press.

Hayami, Yujiro and Yamada, Saburo (1968) 'Technological Progress in Agriculture', in Lawrence R. Klein and Kazushi Ohkawa, (eds), *Economic Growth; The Japanese Experience*, Homewood, Ill., Irwin, pp. 135–61.

Hemmi, Kenzo (1970) *Nogyo* (Agriculture) (Tokyo: Chikuma Shobo).

Hemmi, Kenzo (1982) 'Agriculture and Politics in Japan', in Emery N. Castle and Kenzo Hemmi (eds), *US–Japanese Agricultural Trade Relations*, Baltimore and London, Johns Hopkins University Press, pp. 219–74.

Hewes, Lawrence I. (1950) *Japanese Land Reform*, Tokyo, General Headquarters of the Supreme Commander of Allied Powers.

Hirschman, Albert O. (1970) *Exit, Voice and Loyalty*, Cambridge, Harvard University Press.

Honma, Masayoshi (1987) *Nihon no Nogyo to Zaisei Kozo* (Agriculture and the Structure of Fiscal Policy in Japan), Tokyo, The Forum for Policy Innovation.

Honma, Masayoshi and Hayami, Yujiro (1986) 'Structure of Agricultural Protection in Industrial Countries', *Journal of International Economics*, 20, 2/1, February, pp. 115–29.

Imamuro, Naraumo (1978) *Hajokin to Nogyo* (Subsidies and Agriculture) (Tokyo, Ie no Hekari Kyokai).

Ishi, Mitsuhiro (1981) 'Kazei Shotoku Hosokuritsu no Gyoshukan Kakusa'

(Inter-occupational Differences in the Rates of Capturing Taxable Income), *Kikan Gendai Keizai*, 42, pp. 72–83.

Japan Ministry of Agriculture, Forestry and Fishery (JMAFF) (1984) *Nosei no Kadai to Tenkai Hoko* (Problems and Direction of Agricultural Policy), mimeo.

Johnson, D. Gale (1973) *World Agriculture in Disarray*, London, Macmillan and Fontana; New York, St. Martin's Press.

Johnson, D. Gale, Hemmi, Kenzo, and Lardinois, Pierre (1985) *Agricultural Policy and Trade: Adjusting Domestic Programs in an International Perspective*, A Report to The Trilateral Commission 29, New York, New York Press.

Kajii, Isao (1985) *Nogyo Keiei Kaizen ni Kansuru Shoseisaku no Tenkai* (Development of Policies for the Improvement of Farm Management), in Natsuki Kanazawa (ed.), *Nogyo Keiei to Seisaku* (Farm Management and Policy), Tokyo, Chikusha.

Kawagoe, Toshihiko, Hayami, Yujiro and Ruttan, Vernon W. (1985) 'The Inter-country Agricultural Production Function and Productivity Differences Among Countries', *Journal of Developing Economics*, 19, 1–2, September–October, pp. 114–32.

Kawagoe, Toshihiko, Otsuka, Keijiro and Hayami, Yujiro (1986) 'Induced Bias of Technical Change in Agriculture: The United States and Japan, 1880–1980', *Journal of Political Economy*, 94, 3, June, pp. 523–44.

Kawano, Shigeto (1969) 'Effects of the Land Reform on Consumption and Investment of Farmers', in Kazushi Ohkawa, Bruce F. Johnston and Hiromitsu Kaneda, (eds), *Agriculture and Economic Growth; Japan's Experience*, Tokyo: University of Tokyo Press, pp. 374–97.

Kindleberger, Charles P. (1951) 'Group Behavior and International Trade', *Journal of Political Economy*, 59, 1, February, pp. 30–46.

Krueger, Anne O. (1974) 'The Political Economy of the Rent-Seeking Society', *American Economic Review*, 64, 3, June, pp. 291–303.

Kuroda, Yoshimi and Yoshida, Taiji (1981) 'Production Behavior and Technology of the Farm Household and Marginal Principles in Postwar Japan', *Keizai Kenkyu* (Economic Review), 32, 2, April, pp. 128–41.

Kuznets, Simon (1966) *Modern Economic Growth*, New Haven and London, Yale University Press.

Longworth, John W. (1983) *Beef in Japan: Politics, Production, Marketing and Trade*, St Lucia, London and New York, University of Queensland Press.

Minami, Ryoshin (1986) *The Economic Development of Japan: A Quantitative Analysis*, London: Macmillan.

Mochida, Keizo (1954) '*Shokuryo Seisaku no Seiritsu Katei: Shokuryo Mondai o Meguru Jinushi to Shihon*' (The Process of Formation of Food Policies: Landlords and Capitalists around the Food Problem), *Nogyo Sogo Kenkyu*, 8, 2, April, pp. 197–250.

Namiki, Shokichi (1981) '*Kihonho igo no Nogyo Rodoryoko To Nogyo Seisan* (Agricultural Labor Force and Agricultural Production Since the Basic Law)', *Nogyo Kinyu* (Agricultural Finance), 30, 6, June, pp. 2–9.

Ogura, Takekazu (ed.) (1963) *Agricultural Development of Modern Japan*, Tokyo, Fuji Publishing Co.

Ohkawa, Kazushi (1972) *Differential Structure and Agriculture*, Tokyo, Kinokuniya.

Olson, Mancur (1965) *The Logic of Collective Action*, Cambridge, Harvard University Press.

Olson, Mancur (1985) 'Space, Agriculture, and Organization', *American Journal of Agricultural Economics*, 67, 5, December, pp. 928–37.

Otsuka, Keijiro and Hayami, Yujiro (1985) 'Goals and Consequences of Rice Policy in Japan', *American Journal of Agricultural Economics*, 67, 3, August, p. 529–330.

Ouchi, Tsutomu (1960) *Nogyo Shi* (History of Agriculture), Tokyo, Toyo Keizai Shimposha.

Research Committee on the Basic Problem of Agriculture and Forestry (Noringyo Kihon Mondai Chosakai Jimukyoku) (1960) *Nogyo no Kihon Mondai to Kihontaisaku* (The Basic Problems and Policies of Agriculture), Tokyo: Norin Tokei Kyokai.

Reich, Michael R., Endo, Yasuo and Timmer, Peter C. (1986) 'Agriculture: The Political Economy of Structural Change', in Thomas K. McGraw (ed.), *America versus Japan*, Boston, Harvard University Press, pp. 151–92.

Ricardo, David (1951; original publication 1817) *On the Principle of Political Economy and Taxation*, 3rd edn, Piero Sraffa (ed.), Cambridge, Cambridge University Press.

Sanderson, Fred (1983) *Agricultural Protectionism: Japan, United States and the European Community*, Washington, DC., Japan Economic Institute.

Schultz, Theodore W. (1953) *The Economic Organization of Agriculture*, New York, MacGraw Hill.

Schultz, Theodore W. (1978) 'On Economics and Politics of Agriculture', in Theodore W. Schultz (ed.), *Distortions of Agricultural Incentives*, Bloomington and London, Indiana University Press, pp. 3–23.

Smith, Thomas C. (1959) *The Agrarian Origins of Modern Japan*, Stanford, Stanford University Press.

Stigler, George J. (1971) 'The Theory of Economic Regulation', *Bell Journal of Economics and Management Science*, 2, 1, Spring, pp. 3–21.

Tobata, Seiichi (1956) '*Nihon Nogyo no Ninaite*', (Carriers of Japanese Agriculture), in vol. 9, *Nihon Nogyo Hattatsu Shi* (History of Japanese Agricultural Development), Tokyo, Chuokoronsha, pp. 561–604.

Tollison, Robert D. (1982) 'Rent Seeking: A Survey', *Kyklos*, 35, 4, pp. 574–601.

Tullock, Gordon (1967) 'The Welfare Costs of Tariffs, Monopolies and Theft', *Western Economic Journal*, 5, 3, June, pp. 224–32.

World Bank (1986) *World Development Report 1986*, Oxford, Oxford University Press.

Index

acreage control 63–6, 69–71, 110, 120, 133
Agricultural Act (UK) 77
'agricultural adjustment problem' 14–17, 23, 24, 37, 42–51, 61, 80
agricultural associations 56. *See also* agricultural cooperatives
Agricultural Basic Law 17, 47–8, 74, 77–9, 84, 86, 90, 95
agricultural colleges 31
Agricultural Cooperative Law (1947) 46
agricultural cooperatives: post-war reorganization of 45–7; credit business undertaken by 46, 47, 58, 59, 94; purchasing and marketing 46, 64, 71, 90, 91, 119–21; use of machinery 98, 103–5
Agricultural Development Act (1975) 86
agricultural discussion societies 32
Agricultural Land Commissions 44
Agricultural Land Law 45, 80, 85–8, 95, 114
Agricultural Society (*Nogyokai*) 46
Agricultural Structure-Improvement Programme 80
agriculture: labour productivity in 11–13, 20, 22, 28, 29, 37, 75, 76; post war recovery 43–4, 47; growth rate 26, 28, 29, 112–13; national budget for 57. *See also* farming; farms; labour force
Agriculture and Forestry, Ministry of *see* Japan Ministry
Agriculture, Forestry and Fisheries Credit Corporation 47, 58
Anderson, Kym 127
Anti-Trust Laws 121
Arable Land Replotment Law 33
arare 72, 127
armaments, renunciation of 123
Association of Southeast Asian Nations (ASEAN) 1, 10
Australia 5, 7, 11

Bale, Malcolm D. 13
bananas 52, 54
Bank of Agriculture and Forestry 46
Banks of Agriculture and Industry 33, 34
barley 7, 9, 50, 52, 53, 56, 61–2, 66, 71, 116
beans, import quotas 53, 118
beef 1, 7, 9, 10, 52, 53, 73, 117–19
Bhagwati, Jagdish N. 126
biscuits 127
black market 43, 46, 65
border protection 52–6
Brazil, incomes in 19
Britain, nineteenth-century 35, 39, 41
buffer stocks 56
butter, import control 53
'by products' of NOKYO 94

Canada 7, 116
canals 59
capital formation, ratio of subsidies to 59, 60
capital-intensive products 10, 15, 36, 37, 50, 89

cattle, imports of 55
Central Bank of Agriculture and Forestry (CHUKIN) 46
Central Union of Agricultural Cooperatives (ZENCHU) 46
cheese: import quotas 53; tariffs 55; consumption 116
chicken 7, 9, 10
China 10, 42, 43
citrus fruits *see* grapefruit; lemons; oranges; tangerines
citrus-grower cooperatives 46
coal: import quota 52, 53; government investment in mining 44
combines 96, 97, 104
Common Agricultural Policy of EC 4
comparative advantage, changes in 10–14, 16, 22, 23, 34, 59, 73, 74, 112–13, 121
competition *see* market mechanism
contract work 98, 103–5
cooperatives *see* agricultural cooperatives
core farms 82, 83, 89, 91, 105, 106
corn 7, 9, 55
cost of living 15, 19, 24, 36. *See also* 'food problem'
cream, import quotas 53
credit: business undertaken by cooperatives 46, 47, 58, 59, 94; loans to farmers 33, 34, 38, 47, 58–9
crops: mix of 76. *See also* plant improvement and culture
cultivation rights 113–14

dairy products: protection 10, 53; price support 56; cooperatives 46; output 89; consumption 116, 117. *See also* butter; cheese; cream; milk
deficiency payments 56–7, 119
deflation, effects of 25–6, 37
'democratizing' reforms 44
Denmark 6, 115
Differential Production Scheme 44
ditches 59
Dore, Ronald P. 44
drainage 32, 33, 43, 59
drought (1939) 43
dumping of surpluses 55–6

education of farmers 31, 73, 90, 115, 124
eggs 7, 9, 117
electoral districts 91
Embree, John F. 94
Endo, Yasuo 10
environmental conservation 5, 114, 124–5
European Economic Community: trade policies 2; Common Agricultural Policy 4; protection 5, 6, 10; friction with Japan 10; tariffs 54; subsidies 59, 60
Ever-Normal Granary Plan 38
exchange rates 5, 13, 25–6, 37
extension systems 31, 43, 90, 124

farm households *see* farms, family
Farmland Adjustment Law 44
Farmland Price Control Order 44
Farmland Utilization Promotion Act 87, 95, 114
'farm problem' 14, 15
farm workers *see* labour force; part-time farming
farms, family 26, 45, 48, 76–9, 81, 83–5, 90–93, 98–100, 106–7, 114. *See also* off-farm employment; part-time farming
farms, size of 17, 26, 27, 45, 48, 59, 73–86, 88, 89, 100–5, 114. *See also* scale, economies of; 'viable farm units'
feed grains 2–4, 8, 10, 55, 62, 63, 114–15, 122
female workers 81, 84–5, 95
fertilizer 32, 34, 43, 44, 46, 70, 75, 120, 125, 126
Fertilizer Price Stabilization Act 125, 126
feudalism 25, 31–3

fish 53, 116
fishmeal, tariffs 55
flour, import quotas 53
Food Agency 46, 56, 62, 64, 69, 71
food consumption: factors in 20; Engel coefficient 15, 19, 20; expenditure on 12, 15, 22; changes in 14, 37, 116, 117
Food Control Act (1942) 43, 61
Food Control Special Account 57, 62–4, 66, 115
food control system (*Shokkan Seido*) 4, 43–4, 64; cost of (*Shokuryo Kanrihi*) 63, 66
Food Emergency Measure Act (1946) 43, 44
food grains 8, 10, 117
food problem 15, 16, 19, 23–36, 41–3, 76
food security *see* self-sufficiency
forage crops 116, 122
France 6, 11, 41, 52, 60, 75–7, 115, 117
'free-market rice' (*Jiyu Mai*) 65
fruit 47, 53, 117
Fukuda, Tokuzo 35

gardens 114
General Agreement on Tariffs and Trade (GATT) 2, 52, 54, 118
'general cooperatives' (*Sogo Nokyo*) 46
Germany 6, 11, 39, 41, 60, 75–7, 114, 115, 117
Gerschrenkron, Alexander 12, 41, 127
gono class of farmers
'government-rice channel' (*Seifu Mai*) 64
grains 8, 10, 53, 122. *See also* feed grains; food grains
grapefruit 52, 54, 118
grapes 30

Hayami, Yujiro 12, 18, 22, 26, 29, 30, 33, 36, 37, 67, 76, 97, 119, 127
Hemmi, Kenzo 8, 49, 54, 72
Hewes, Lawrence I. 44
hinshukokankai (seed exchange society) 32
Hirschman, Albert O. 72, 127
Hokkaido 83–4, 89
Hokuriku 100, 102
Honma, Masayoshi 61

Imperial Agricultural Society 35, 37
imports 2–4, 12, 20, 52–6, 72; quotas 52–5, 71, 118, 119, 126. *See also* rice, imports of
incomes: relation to food consumption and prices 12, 14, 15, 22, 37, 41, 70; disparity between farm and non-farm 12, 14, 22, 24, 41, 42, 47, 69, 74, 77, 80, 82, 89, 90, 92–3, 113, 123; relation to industrial structure 18–21; relation to farm size 100. *See also* off-farm employment; poverty; wages
India 11, 12
Indonesia 19, 56
industry: recovery, productivity and growth 11–13, 16, 20, 22, 24, 25, 28, 29, 30, 34, 36–7, 39, 41, 51; incomes in 16, 18–21, 24, 49–50; relation to food production 30, 34, 39, 43, 44, 47; manufacturing 20, 22, 23, 41
inflation 45
infrastructure *see* land infrastructure
innovation 32, 32, 34
input as factor in productivity 28–30, 59. *See also* fertilizer
insurance undertaken by cooperatives 46, 94
iron and steel industries 44
irrigation 32–4, 36, 43, 59, 70, 71
Ishi, Mitsuhiro 60
Italy 6, 41, 75, 117

Japan Hypothec Bank 33
Japan Ministry of Agriculture

Forestry and Fisheries (JMAFF) 46, 58, 72, 114, 125, 127. *See also* Food Agency
Japan Tobacco Corporation 52
Jishu Ryutsu Mai (voluntary-rice channel) 64, 65
Jiyu Mai ('free-market rice') 65
Johnson, D. Gale 8, 13, 54
Junkers 41, 126–7

Kajii, Isao 86
Kanno Seisaku (Policies to Encourage Agricultural Production) 30
Kawagoe, Toshihiko, 30, 97
Kawano, Shigeto 45
Keisha Seisanhoshiki (Differential Production Scheme) 44
Kikuchi, Masao 33
Kinai Branch experimental station 36
Kindleberger, Charles P. 41
Kinki, experimentation in 32
Kome Sodo (Rice Riot) 35, 36
konnyaku roots 53
Korea 2, 5, 6, 8, 11, 12, 19, 35–7, 43, 44, 56
Krueger, Anne O. 126
Kuroda, Yoshimi 97
Kuznets, Simon 30
Kyushu, North, experimentation in 32

labour productivity 11–13, 20, 22, 28, 29, 37, 75, 76
labour force, agricultural: 19, 20, 78–9, 81, 82; movements in and out of farming 14, 37, 47, 49, 51, 108–11; composition of 105–7; elderly 84, 90, 91, 95, 99, 105, 108–10; female 84–5, 95. *See also* part-time farming
labour-intensive products 10
labour-saving technology 17, 26, 73, 113
land: scarcity of 10; conversion to non-agricultural use 83; price and rates of return 86–7, 99; land-improvement projects 33, 34; land infrastructure 58–9, 72, 77, 80, 115; land-labour ratio 28, 29, 75; land productivity 28, 29, 32, 34, 75, 76; land reform (1946–50) 44–7; land-saving technology 26, 30, 32, 75–7, 97, 99, 113, 126; land tenure 17, 25, 26, 31, 38, 85–8, 108, 114; leasing of land 86–8, 95, 102–5, 113. *See also* farms, size of; landlords; pasture; tenant farmers
Land Tax Revision 25, 32
landlords 25, 26, 32–4, 38, 39, 42, 44–5, 49, 85–7, 102
Landwirtschaft Recht, Das 77
Lardinois, Pierre 54
leather, tariffs on 52, 55
lemons 52, 54
levies on imports 54–6
Liberal Democrat Party 49, 72, 125, 127
life-expectancy 84–5, 105
livestock 10, 30, 47, 97, 115, 116, 122. *See also* beef; cattle; chicken; dairy products; eggs; pigs; poultry
Livestock Industry Promotion Corporation 52, 56
loans *see* credit
Loi d'orientation agricole 77
Longworth, John W. 119
Lutz, Ernest 13

maize 8
malt 55
managerial ability 99, 110
manufacturing *see* industry
market mechanism 119–21, 125–7
marketing: as proportion of food expenditure 15, 19, 20; subsidized 80; by cooperatives 46, 64, 71, 90, 91, 94, 119–21; cost of 67
Matsukata Deflation 25–6
meat 53, 56, 116, 117

mechanisation: and farm size 76, 95–100, 102, 105; and oil supplies 122
Meiji period 16–19, 22, 24, 25, 30–34, 36, 38, 39
mikan 55, 118, 126
milk 7, 9, 52, 53, 56, 116, 117
mochi 127
Mochida, Keizo 35
modernization, subsidization for 80
monopolies of government agencies 52
mugi 66

Namiki, Shokichi 107, 108
National Federation of Agricultural Cooperatives (ZENNOH) 46
National Insurance Agricultural Cooperative Federation 46
Netherlands 6, 114–15
New Village Construction Programme 47
New Zealand 7, 10
nodankai (agricultural discussion society) 32
Nogyokai (Agricultural Society) 46
NOKYO (*Nogyo Kyodo Kumiai*) 46, 47, 49, 50, 71, 72, 91, 94, 119–21, 125–8
nominal rate of protection 4–11, 41, 42

oats 7, 9, 55
off-farm employment and income 22, 49, 81–4, 90–94, 106, 107, 123, 124, 126
Ogura, Takekazu 26, 44, 47
Ohkawa, Kazushi 37
oil supplies 11, 64, 73, 108, 122
olives 30
Olson, Mancur 94
oranges 1, 10, 53, 55, 118. *See also* tangerines
Otsuka, Keijiro 30, 67
Ouchi, Tsutomu 26, 34, 38
Owner-Farmer Establishment Special Measure Law (1946) 44

paddy fields: withdrawal of 63, 64, 66–8, 71; rate of return on 86, 87; subsidies for diversion of 120; environmental value 124
part-time farming 17, 81–95, 106, 107, 113, 115, 119–20, 125
pasture, improvement of 47
peanuts 53, 118
pensions 124
Philippines 11, 12, 19
pigs 10, 89. *See also* pork
pineapples 53
plant culture and improvement 30–34, 36
plywood 10
political relation with protection 46, 48, 49, 51, 64, 70, 72, 91, 125–7, 129–31
population density 14, 15, 29, 33, 83
pork 7, 9, 54–5
potatoes 44, 56
poultry 10, 89. *See also* chicken
poverty, rural 12, 14, 15, 17, 22, 24, 36–9, 41, 42, 58. *See also* incomes
power tillers 17, 84, 95–7
Price Stabilization Programmes 56–7
price support 56–7. *See also under* rice
processed foods 72, 127
processing of food as share of food expenditure 15, 19, 20
Production Cost and Income Compensation Formula 48, 62, 69
protection, agricultural: increase in 1–2, 4–12; resistance to 15, 16, 24, 112; foreign objections to 1, 2, 10, 13, 16, 72, 73, 112, 116, 118, 127, 128; instruments of 51–61; effects of 61–73; means of reducing 121–8; political and sociological

reasons for 12, 15–17, 23, 34, 38–9, 41, 74, 91, 124–31. *See also* border protection; import quotas; levies; nominal rate of protection; rice, price support; subsidies; tariffs; taxation
protection, industrial 22, 23
protein consumption 116, 117
pulses 2–4

quarantine: for livestock 55; for plants 118

rapeseed 56
rates of return on land 87
rationing 43, 61
raw foodstuffs 15, 19
raw materials, imports 10, 72
reapers 96
Reich, Michael R. 10
Rent Control Order (1939) 44
rents 26, 38, 44–5, 72, 85, 86, 99–100, 114
research 31, 36, 43, 70, 73, 90, 113, 115, 124
Ricardo, David 15, 16, 35
rice: nominal rate of protection of 7–9, 41, 42; consumption of 8, 10, 47, 49–50, 63, 70, 116, 117; shortages 43, 44; imports 34–5, 37, 38, 41, 52, 62, 66–7, 127; self-sufficiency 3, 10, 34–6, 43, 47; surplus 25, 26, 39, 48, 49, 55, 62–5, 68–71, 100, 101, 132–4; exports 55–6, 62; food control system and price support 4, 10, 17, 37, 38, 43, 48–9, 56, 61–71, 94, 110, 115–16, 119, 120, 124; improvements in culture and production 31, 33, 35–6, 63, 115; size of farms and mechinisation 73, 84, 89, 90, 96–8, 100–101, 113; marketing 46, 71, 90, 94, 119; effects of rice policy 132–4. *See also* paddy fields
Rice Control Law (1933) 38
Rice Distribution Control Act (1939) 43, 44
Rice Law (1921) 25, 31, 32, 38
Rice Production Development Programme 36
Rice Riot (1918) 35, 36, 41
Rikuu Branch experimental station 36
rono 31, 32
Russo-Japanese war 35
Ruttan, Vernon W. 12, 22, 29, 37, 76, 97
rye 7, 9

Sanderson, Fred 119
Sanmai Zoshoku Keikaku (Rice Production Development Programme) 36
scale, economies of 76, 96–100, 105, 120. *See also* farms, size of
school-leavers 78–9, 109
Schultz, Theodore W. 13, 14, 15, 30
seaweed 53
seed exchange societies 32
self-sufficiency 2–4, 10, 11, 25, 34–6, 38, 43, 47, 70, 121–3
Senmon NOKYO 46
share tenancy 114
shellfish 53
Shokkan Seido (Food Control system) 4, 43–4, 64
Shokuryo Kanrihi (food control cost) 66
silk 10, 52, 55, 56
Silk and Sugar Price Stabilisation Corporation 52, 55, 56
Smith, Thomas C. 26
social policies in agriculture 38–9, 42, 113
socialism, fear of 39, 50
Sogo NOKYO 46
sorghum 55
Soviet Union 2
soybean 2, 10, 56, 70, 116, 121
spaghetti 72, 127

sprayers 96
starch 53, 55, 56
state-trading 52, 54, 56, 62, 71, 118
Stigler, George J. 126
storage cost 62, 70
subsidies 34, 38–40, 57–61, 66, 67, 72, 80, 118, 120, 126
Subsidization for Agricultural Modernization Funds, Act of 80
subsistence farming 124–5
sugar 7, 9, 53, 55, 56
surpluses 8, 14, 47, 55–6, 100–101. *See also under* rice
Sweden 5, 7
Switzerland 5, 7

Taiwan 2, 5, 6, 8, 12, 35–7
tangerines 55, 118, 126
tariffs 34–5, 41, 52, 54–5
taxation: of agricultural and non-agricultural sectors 39, 40, 60–61, 110; revenues 64, 69, 73. *See also* Land Tax Revision
technology: borrowing of 12, 13, 22, 30, 112, 113; imports of 25, 30. *See also* labour-saving; land-saving
tenant farmers 38, 44–5, 85, 86, 99–100, 102–5, 114
Thailand 19, 33, 127
threshers 95, 96
Timmer, Peter C. 10
tobacco 52
Tobata, Seiichi 34, 35, 123
Tohoku 100, 102
Tokugawa period 25, 31–3
Tokyo Chamber of Commerce 35
Tollison, Robert 126
tomatoes 10, 53
tractors 17, 96–8, 104
trade policies 2, 116–19, 122, 124. *See also* market mechanism; protection
transplanters 96, 97
Tullock, Gordon 126

United Kingdom 6, 11, 60, 75, 77, 117
United States of America 2, 5, 7, 10, 11, 29, 43, 47, 54, 70, 75, 117, 120, 121

vegetables 10, 47, 89, 117
veteran farmers' techniques 31, 32
'viable farm units' 76–8, 80, 81, 84, 88–91, 95, 99, 103, 108, 110
village communities, social structure of 33, 94, 126
village development plans 47, 80; land-use leases 86, 87, 95, 105
'voluntary-rice channel' (*Jishu Ryutsu Mai*) 64–6

wages: effect of protection on 15; and cost of living 16, 19, 24, 36, 42; and cost of production 76, 97; in large- and small-scale industries 37; in rice-price calculations 48, 69. *See also* income
water-pumps 95
wheat 2–4, 7–9, 44, 56, 61–2, 66, 71, 114, 127
wine 30

Yamada, Saburo 36
Yokōi, Jikei 35
Yoshida, Taiji 97

ZENCHU (Central Union of Agricultural Cooperatives) 46
ZENKYOREN (National Insurance Agricultural Cooperative Federation) 46
ZENNOH (National Federation of Agricultural Cooperatives) 46

